1+X 证书制度试点培训用书

Web 前端开发实训案例教程（Java）

北京新奥时代科技有限责任公司　组　编

电子工业出版社
Publishing House of Electronics Industry
北京·BEIJING

内 容 简 介

本书是与"Web 前端开发职业技能等级标准"配套的教材,主要应用 Java 搭建动态网站,书中所有代码均在主流浏览器中运行通过。

全书分为两大部分:第一部分为 Java 动态网站搭建基础,对 SSM 框架中的各知识单元的要点进行阐述,并辅以代码示例进行诠释和应用;第二部分为实践,可以对应课程实验或综合实践,采用技术专题进行重要知识单元训练,参考企业项目开发过程和标准,针对不同的知识单元设计了项目,重点训练每一个知识单元内容。

本书适合作为"Web 前端开发职业技能等级标准"的教学参考用书,也适合作为对 Web 前端开发感兴趣的读者的编程指导用书。

未经许可,不得以任何方式复制或抄袭本书之部分或全部内容。
版权所有,侵权必究。

图书在版编目(CIP)数据

Web 前端开发实训案例教程:Java / 北京新奥时代科技有限责任公司组编. —北京:电子工业出版社,2022.3

ISBN 978-7-121-43097-8

Ⅰ.①W… Ⅱ.①北… Ⅲ.①网页制作工具—教材②JAVA 语言—程序设计—教材 Ⅳ.①TP393.092.2 ②TP312.8

中国版本图书馆 CIP 数据核字(2022)第 042726 号

责任编辑:胡辛征
印　　刷:涿州市京南印刷厂
装　　订:涿州市京南印刷厂
出版发行:电子工业出版社
　　　　　北京市海淀区万寿路 173 信箱　　邮编:100036
开　　本:787×1092　1/16　印张:13.5　字数:342.2 千字
版　　次:2022 年 3 月第 1 版
印　　次:2022 年 3 月第 1 次印刷
定　　价:49.80 元

凡所购买电子工业出版社图书有缺损问题,请向购买书店调换。若书店售缺,请与本社发行部联系,联系及邮购电话:(010) 88254888,88258888。
质量投诉请发邮件至 zlts@phei.com.cn,盗版侵权举报请发邮件至 dbqq@phei.com.cn。
本书咨询联系方式:(010) 88254361,hxz@phei.com.cn。

前 言

为帮助读者学习和掌握"Web 前端开发职业技能等级标准"标准（中级）中涵盖的职业技能要求，工业和信息化部教育与考试中心 1+X 项目组组织企业工程技术人员编写本书。本书按照标准中涉及的核心技能要求精心设计"Java 动态网站搭建"重要知识单元和配套技术专题，这些技术专题全部按照企业项目开发思路进行分析设计和实施。

本书结合高校计算机相关专业 Web 前端开发方向课程体系、企业 Web 前端开发岗位能力模型和"Web 前端开发职业技能等级标准"，形成 Web 前端开发三位一体知识地图，以实践能力为导向，以企业真实应用为目标，遵循企业软件工程标准和技术，以任务为驱动，对 Java 编程基础、类和对象、Servlet、JSP、数据库编程（JDBC）、SSM 框架、Bootstrap、AJAX、RESTful API 等 Web 前端开发中的重要知识单元，结合实际案例和应用环境进行分析和设计，并对每个重要的知识单元进行详细的讲解，力求帮助读者真正掌握这些知识在实际场景中的应用。

全书分为 Java 动态网站搭建基础和实践两部分，共 15 章，其中：

第一部分是 SSM 框架，包括现在主流三大框架，重点讲解三大框架的核心知识及应用。

第二部分是专题训练（实验），以任务为驱动，采用迭代的思路进行开发。针对 Bootstrap、MySQL 数据库操作、Java 类和对象、JSP 和 Servlet、数据库编程、SSM 框架编程、RESTful API、AJAX 和 JSON 等核心知识单元设计技术专题，每个技术专题针对一次实验项目进行训练，内容包括实验目标、实验任务、设计思路和实验实施（跟我做），最大程度覆盖 Web 前端开发中级实践内容。

参加本书编写工作的人员有谭志彬、龚玉涵、张晋华、马庆槐、郑婕、马玲等。

由于编者水平有限，书中难免存在不足之处，敬请读者批评和指正。

编　者

目录

第1章　SSM 框架 ··················· 1
　1.1　SSM 框架简介 ··············· 1
　1.2　持久层框架 MyBatis ········· 3
　　1.2.1　MyBatis 简介 ············ 3
　　1.2.2　MyBatis 配置 ············ 3
　　1.2.3　XML 映射文件 ··········· 6
　　1.2.4　动态 SQL ··············· 10
　1.3　Spring 框架 ················ 13
　　1.3.1　Spring IoC ·············· 13
　　1.3.2　容器 ··················· 13
　　1.3.3　Spring Bean ············ 14
　1.4　SpringMVC 框架 ··········· 16
　　1.4.1　SpringMVC 开发流程 ··· 16
　　1.4.2　分派器 ················· 16
　　1.4.3　控制器 ················· 17
　　1.4.4　视图解析 ··············· 19
　　1.4.5　ModelAndView 类 ······· 20
第2章　Bootstrap 页面开发：
　　　　注册页面 ················· 21
　2.1　实验目标 ·················· 21
　2.2　实验任务 ·················· 21
　2.3　设计思路 ·················· 22
　2.4　实验实施（跟我做） ······· 23
第3章　响应式网页开发：
　　　　分类信息页面 ············· 29
　3.1　实验目标 ·················· 29
　3.2　实验任务 ·················· 29
　3.3　设计思路 ·················· 30
　3.4　实验实施（跟我做） ······· 34
第4章　创建第一个 Java 程序 ····· 40
　4.1　实验目标 ·················· 40
　4.2　实验任务 ·················· 40
　4.3　设计思路 ·················· 40
　4.4　实验实施（跟我做） ······· 41
第5章　类和对象：日期计算器 ···· 54
　5.1　实验目标 ·················· 54

　5.2　实验任务 ·················· 54
　5.3　设计思路 ·················· 55
　5.4　实验实施（跟我做） ······· 56
第6章　Java Web 编程（JSP）：
　　　　在线投票 ················· 63
　6.1　实验目标 ·················· 63
　6.2　实验任务 ·················· 63
　6.3　设计思路 ·················· 64
　6.4　实验实施（跟我做） ······· 65
第7章　Java Web 编程（JSP+Servlet）：
　　　　购物车 ··················· 75
　7.1　实验目标 ·················· 75
　7.2　实验任务 ·················· 75
　7.3　设计思路 ·················· 77
　7.4　实验实施（跟我做） ······· 79
第8章　MySQL 数据库：
　　　　MySQL 基本操作 ·········· 93
　8.1　实验目标 ·················· 93
　8.2　实验任务 ·················· 93
　8.3　设计思路 ·················· 93
　8.4　实验实施（跟我做） ······· 94
第9章　MySQL 数据库：
　　　　试题信息管理 ············· 98
　9.1　实验目标 ·················· 98
　9.2　实验任务 ·················· 98
　9.3　设计思路 ················· 100
　9.4　实验实施（跟我做） ······ 102
第10章　数据库编程（JDBC）：
　　　　 学生成绩管理 ··········· 112
　10.1　实验目标 ················ 112
　10.2　实验任务 ················ 112
　10.3　设计思路 ················ 114
　10.4　实验实施（跟我做） ····· 117

第 11 章	SSM 框架：		13.3	设计思路	169
	第一个 SSM 程序	136	13.4	实验实施（跟我做）	173
11.1	实验目标	136	第 14 章	RESTful API 规范：	
11.2	实验任务	136		视频列表	187
11.3	设计思路	137	14.1	实验目标	187
11.4	实验实施（跟我做）	138	14.2	实验任务	187
第 12 章	SSM 框架：在线答题	155	14.3	设计思路	188
12.1	实验目标	155	14.4	实验实施（跟我做）	190
12.2	实验任务	155	第 15 章	AJAX：天气预报	199
12.3	设计思路	156	15.1	实验目标	199
12.4	实验实施（跟我做）	157	15.2	实验任务	199
第 13 章	SSM 框架：个人博客	167	15.3	设计思路	200
13.1	实验目标	167	15.4	实验实施（跟我做）	201
13.2	实验任务	167	参考文献		208

第 1 章 SSM 框架

1.1 SSM 框架简介

Java SSM 框架是 Spring+SpringMVC+MyBatis 的简称,由 Spring 和 MyBatis 两个开源框架整合而成(SpringMVC 是 Spring 中的一部分),相比于 SSH(Struts+Spring+Hibernate),SSM 更加轻量化和灵活,是目前业界主流的 Java Web 开发框架。

1)SSM 框架包

Spring、MyBatis 和 SpringMVC 包主要涉及如图 1-1 所示的内容。

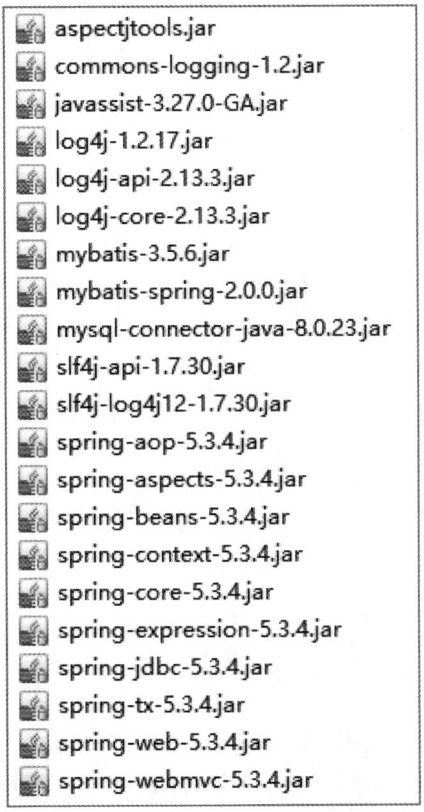

```
aspectjtools.jar
commons-logging-1.2.jar
javassist-3.27.0-GA.jar
log4j-1.2.17.jar
log4j-api-2.13.3.jar
log4j-core-2.13.3.jar
mybatis-3.5.6.jar
mybatis-spring-2.0.0.jar
mysql-connector-java-8.0.23.jar
slf4j-api-1.7.30.jar
slf4j-log4j12-1.7.30.jar
spring-aop-5.3.4.jar
spring-aspects-5.3.4.jar
spring-beans-5.3.4.jar
spring-context-5.3.4.jar
spring-core-5.3.4.jar
spring-expression-5.3.4.jar
spring-jdbc-5.3.4.jar
spring-tx-5.3.4.jar
spring-web-5.3.4.jar
spring-webmvc-5.3.4.jar
```

图 1-1

2）SSM 环境

Java Development Kit（JDK）是 Sun 公司针对 Java 开发人员发布的免费软件开发工具包（Software Development Kit，SDK）。自从 Java 推出以来，JDK 已经成为使用最广泛的 Java SDK。作为 Java 语言的 SDK，普通用户并不需要安装 JDK 来运行 Java 程序，而只需安装 JRE（Java Runtime Environment）即可；而程序开发者则必须安装 JDK 来编译、调试程序。

Eclipse 是 Java 集成开发工具，是一个开放源代码的、基于 Java 的可扩展开发平台。就其本身而言，它只是一个框架和一组服务，用于通过插件组件构建开发环境。幸运的是，Eclipse 附带了一个标准的插件集，包括 JDK。Eclipse 的启动页面如图 1-2 所示。

图 1-2

使用 Eclipse 创建 SSM 工程，在 WEB-INF 文件夹下将 Spring、MyBatis 和 SpringMVC 的 jar 包放到 lib 目录下，如图 1-3 所示。

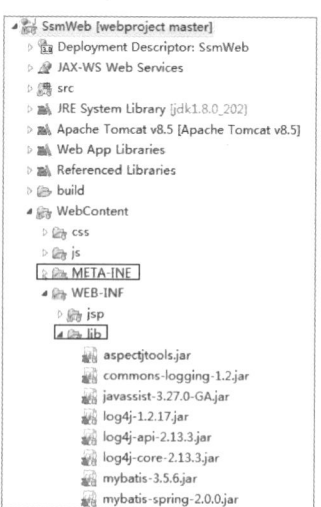

图 1-3

3）SSM 架构集成过程

SSM 架构图（以用户请求为例）如图 1-4 所示，Web 项目启动时，服务器加载 web.xml 配置文件，在 web.xml 文件中创建 Spring 的 DisptchcerServlet 类对象，并加载 springmvc.xml 和 applicationContext.xml 配置文件。其中，applicationContext.xml 文件用来集成 Spring 和 MyBatis，通过 Spring 管理 SqlSessionFactory、mapper 接口。

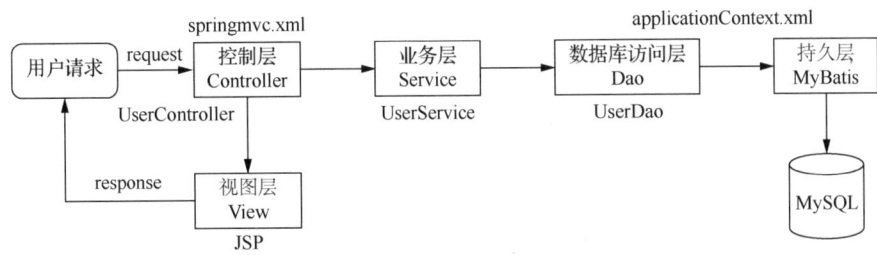

图 1-4

1.2 持久层框架 MyBatis

1.2.1 MyBatis 简介

MyBatis 是一款优秀的持久层框架，它支持自定义 SQL（Structure Query Language，结构查询语言）、存储过程及高级映射。MyBatis 免除了大多数的 JDBC（Java Database Connectivity，Java 数据库互连）代码，以及设置参数和获取结果集的工作。MyBatis 可以通过简单的 XML（Extensible Markup Language，可扩展标记语言）或注解来配置和映射原始类型、接口和 Java POJO（Plain Old Java Objects，普通老式 Java 对象）。

1.2.2 MyBatis 配置

1. properties 属性

这些属性可以在外部进行配置，并可以进行动态替换。

（1）在 src 目录下创建 db.properties 文件，编写数据库配置信息。

```
jdbc.driver=com.mysql.cj.jdbc.Driver
jdbc.url=jdbc:mysql://localhost:3306/blogdb
jdbc.username=root
jdbc.password=12345
```

（2）读取 db.properties 文件设置属性。

```
<!DOCTYPE configuration
  PUBLIC "-//mybatis.org//DTD Config 3.0//EN"
  "http://mybatis.org/dtd/mybatis-3-config.dtd">
<configuration>
  <properties resource="db.properties">
    <property name="driver" value="jdbc.driver"/>
    <property name="url" value="jdbc.user"/>
    <property name="username" value="jdbc.username"/>
    <property name="password" value="jdbc.password"/>
  </properties>
</configuration>
```

设置好的属性可以在整个配置文件中用来替换需要动态配置的属性值。

```
<!DOCTYPE configuration
  PUBLIC "-//mybatis.org//DTD Config 3.0//EN"
  "http://mybatis.org/dtd/mybatis-3-config.dtd">
```

```xml
<configuration>
  <properties resource="db.properties">
    <property name="driver" value="jdbc.driver"/>
    <property name="url" value="jdbc.user"/>
    <property name="username" value="jdbc.username"/>
    <property name="password" value="jdbc.password"/>
  </properties>

  <environments default="development">
    <environment id="development">
      <transactionManager type="JDBC"/>
      <dataSource type="POOLED">
        <property name="driver" value="${driver}"/>
        <property name="url" value="${url}"/>
        <property name="username" value="${username}"/>
        <property name="password" value="${password}"/>
      </dataSource>
    </environment>
  </environments>
</configuration>
```

从 MyBatis 3.4.2 开始，可以为占位符指定一个默认值。例如：

```xml
<dataSource type="POOLED">
  <property name="username" value="${username:root}"/>
  <!-- 如果属性 'username' 没有被配置，则'username' 属性的值将为 'root' -->
</dataSource>
```

2. settings 设置

一个配置完整的 settings 元素的示例如下。

```xml
<settings>
  <setting name="cacheEnabled" value="true"/>
  <setting name="lazyLoadingEnabled" value="true"/>
  <setting name="multipleResultSetsEnabled" value="true"/>
  <setting name="useColumnLabel" value="true"/>
  <setting name="useGeneratedKeys" value="false"/>
  <setting name="autoMappingBehavior" value="PARTIAL"/>
  <setting name="autoMappingUnknownColumnBehavior" value="WARNING"/>
  <setting name="defaultExecutorType" value="SIMPLE"/>
  <setting name="defaultStatementTimeout" value="25"/>
  <setting name="defaultFetchSize" value="100"/>
  <setting name="safeRowBoundsEnabled" value="false"/>
  <setting name="mapUnderscoreToCamelCase" value="false"/>
  <setting name="localCacheScope" value="SESSION"/>
  <setting name="jdbcTypeForNull" value="OTHER"/>
  <setting name="lazyLoadTriggerMethods" value="equals,clone,hashCode,toString"/>
</settings>
```

3. typeAliases 类型别名

类型别名可为 Java 类型设置一个缩写名称。它仅用于 XML 配置，意在降低冗余的全限定类名书写。例如：

```xml
<typeAliases>
  <typeAlias alias="Blog" type="com.blogs.po.Blog"/>
  <typeAlias alias="User" type="com.blogs.po.User"/>
</typeAliases>
```

当这样配置时，Blog 可以用在任何使用 com.blog.model.Blog 的地方。也可以指定一个包名，MyBatis 会在包名下面搜索需要的 JavaBean。例如：

```xml
<typeAliases>
  <package name="com.blogs.po"/>
</typeAliases>
```

每一个在包 co.blogs.po 中的 Java Bean，在没有注解的情况下，会使用 bean 的首字母小写的非限定类名来作为它的别名；若有注解，则别名为其注解值。在 JavaBean 中使用 @Alias( ) 函数添加别名注解。

```java
package com.blogs.model;
import org.apache.ibatis.type.Alias;

@Alias("blog")
public class Blog{
……
}
```

4. environments 环境配置

environments 元素定义了如何配置环境，例如：

```xml
<environments default="development">
  <environment id="development">
    <transactionManager type="JDBC">
      <property name="…" value="…"/>
    </transactionManager>
    <dataSource type="POOLED">
      <property name="driver" value="${driver}"/>
      <property name="url" value="${url}"/>
      <property name="username" value="${username}"/>
      <property name="password" value="${password}"/>
    </dataSource>
  </environment>
</environments>
```

5. transactionManager 事务管理器

在 MyBatis 中有两种类型的事务管理器，即 JDBC 和 MANAGED。

（1）JDBC：直接简单使用 JDBC 的提交和回滚设置。它依赖于从数据源得到的连接来管理事务范围。

（2）MANAGED：它从来不提交或回滚一个连接。而会让容器来管理事务的整个生命周期，默认情况下它会关闭连接。然而一些容器并不希望这样，因此如果需要从连接中停止，则将 closeConnection 属性设置为 false，代码如下：

```xml
<transactionManager type="MANAGED">
  <property name="closeConnection" value="false"/>
</transactionManager>
```

6. mappers 映射器

我们需要告诉 MyBatis 到哪里去找到定义 SQL 映射语句的文件。可以使用相对于类路径的资源引用，或完全限定资源定位符（包括 file:/// 形式的 URL），或类名和包名等。例如：

```xml
<!-- 使用相对于类路径的资源引用 -->
<mappers>
  <mapper resource="org/mybatis/builder/AuthorMapper.xml"/>
  <mapper resource="org/mybatis/builder/BlogMapper.xml"/>
  <mapper resource="org/mybatis/builder/PostMapper.xml"/>
</mappers>

<!-- 使用完全限定资源定位符（URL） -->
<mappers>
  <mapper url="file:///var/mappers/AuthorMapper.xml"/>
  <mapper url="file:///var/mappers/BlogMapper.xml"/>
  <mapper url="file:///var/mappers/PostMapper.xml"/>
</mappers>

<!-- 使用映射器接口实现类的完全限定类名 -->
<mappers>
  <mapper class="org.mybatis.builder.AuthorMapper"/>
  <mapper class="org.mybatis.builder.BlogMapper"/>
  <mapper class="org.mybatis.builder.PostMapper"/>
</mappers>

<!-- 将包内的映射器接口实现全部注册为映射器 -->
<mappers>
  <package name="org.mybatis.builder"/>
</mappers>
```

1.2.3　XML 映射文件

1. select

一个简单查询的 select 元素是非常简单的。例如：

```xml
<?xml version="1.0" encoding="UTF-8"?>
<!DOCTYPE mapper PUBLIC "-//mybatis.org//DTD Mapper 3.0//EN" "http://mybatis.org/dtd/mybatis-3-mapper.dtd">
<mapper namespace="users">
  <!-- 根据 id 查询得到一个 user 对象 -->
  <select id="selectUserById" parameterType="int"
```

```
    resultType="User">
    select *
    from users where id=#{id}
  </select>
</mapper>
```

上述语句的名称为 selectUserById，接收一个 int（或 Integer）类型的参数、User 类型的对象，其中的键是列名，值便是结果行中的对应值。

2. insert、update 和 delete

数据变更语句 insert、update 和 delete 的实现非常接近。例如：

```
<?xml version="1.0" encoding="UTF-8"?>
<!DOCTYPE mapper PUBLIC "-//mybatis.org//DTD Mapper 3.0//EN" "http://mybatis.org/dtd/mybatis-3-mapper.dtd">
<mapper namespace="users">

  <insert id="insertOne" parameterType="User"
    useGeneratedKeys="true" keyProperty="id">
    insert into
    users(account,password)
    values
    (#{account},#{password})
  </insert>

  <update id="updateOne" keyProperty="id" keyColumn="id">
    update
    users set
    account=#{account},
    password=#{password} where id=#{id}
  </update>

  <delete id="deleteById" parameterType="int">
    delete from users where
    id=#{id}
  </delete>

</mapper>
```

3. 参数

之前见到的所有语句都使用了简单的参数形式。但实际上，参数是 MyBatis 非常强大的元素。对于大多数简单的使用场景，不需要使用复杂的参数。例如：

```
<select id="selectUsers" resultType="User">
  select id, username, password
  from users
  where id=#{id}
</select>
```

在下面的 search 语句中，参数有 uid 和 keyword 两个。

```
<select id="search" resultMap="blogMap">
```

```
  select b.*, u.account from blogs b, users u where b.user_id
  =u.id and b.user_id=#{uid} and (b.title like #{keyword} or b.content like
  #{keyword})
</select>
```

调用 search 方法时,可以将传入的参数放到 Map 对象中,然后传入。

```
public List<Blog> searchByKeyword(String keyword, int uid){
  Map<String, Object> params=new HashMap();
  params.put("uid", uid);
  params.put("keyword", '%' + keyword + '%');
  return blogMapper.search(params);
}
```

4. 结果映射

ResultMap 的设计思想是,对简单的语句做到零配置;对于复杂的语句,只需要描述语句之间的关系即可。

简单映射语句没有显式指定 resultMap,例如:

```
<select id="selectUsers" resultType="map">
  select id, username, hashedPassword
  from some_table
  where id=#{id}
</select>
```

结果集的类型由 resultType 属性指定,map 表示是 HashMap 类型,上述语句只是简单地将所有的列映射到 HashMap 的键上,这由 resultType 属性指定。

程序可能会使用 JavaBean 或 POJO 作为领域模型。例如,定义如下 JavaBean:

```
public class User{
  private int id;
  private String username;
  private String hashedPassword;

  public int getId(){
    return id;
  }
  public void setId(int id){
    this.id=id;
  }
  public String getUsername(){
    return username;
  }
  public void setUsername(String username){
    this.username=username;
  }
  public String getHashedPassword(){
    return hashedPassword;
  }
  public void setHashedPassword(String hashedPassword){
    this.hashedPassword=hashedPassword;
```

}
　}

　　基于 JavaBean 的规范，上面这个类有 3 个属性：id、username 和 hashedPassword。这些属性会对应到 select 语句中的列名。这样的一个 JavaBean 可以被映射到 ResultSet，就像映射到 HashMap 一样简单。例如：

```xml
<select id="selectUsers" resultType="User">
 select id, username, hashedPassword
 from some_table
 where id=#{id}
</select>
```

　　如果列名和属性名不能匹配上，则可以通过在 SELECT 语句中设置列别名（这是一个基本的 SQL 特性）来完成匹配。例如：

```xml
<select id="selectUsers" resultType="User">
 select
  user_id as "id",
  user_name as "userName",
  hashed_password as "hashedPassword"
 from some_table
 where id=#{id}
</select>
```

5. id 和 result

定义一个结果映射类型，例如：

```xml
<resultMap type="Blog" id="blogMap">
  <id property="id" column="id"/>
  <result property="title" column="title"/>
  <result property="content" column="content"/>
  <result property="create_time" column="create_time"/>
  <association property="user" javaType="User">
    <id property="id" column="user_id"/>
    <result property="account" column="account"/>
  </association>
</resultMap>
```

　　id 和 result 元素都将一个列的值映射到一个简单数据类型（如 String、int、double、Date 等）的属性或字段。

6. 关联

　　关联元素用来描述"某一个类中的属性为另一个类的对象"这种关系，如博客中有一个用户属性。MyBatis 有两种不同的方式加载关联。

　　（1）嵌套 Select 查询：通过执行另外一个 SQL 映射语句来加载关联的属性。

　　（2）嵌套结果映射：使用嵌套的结果映射来处理连接结果的重复子集。

　　在下面的示例中，Blog 类型中有一个 User 类型，需要指定属性名 user 和属性的 javaType 为"User"类型。通过设置 id 元素来指明该属性对应于表中的字段。MyBatis 在加载时，先读取 Blog 表中 user_id 字段的值，再找到 User 表中 id 与 user_id 值相等的记录，来构建 User 对象。

```xml
<resultMap type="Blog" id="blogMap">
  ……
  <association property="user" javaType="User">
    <id property="id" column="user_id"/>
    <result property="account" column="account"/>
  </association>
</resultMap>
```

7. 集合

集合元素和关联元素相似，一个博客只有一个作者，可以使用关联来表示。但一个博客有很多文章，这时就需要使用集合来表示。如下面代码所示，使用 collection 元素来定义博客中一个文章的集合，使用 id 元素来指明该属性对应于表中的字段。

```xml
<resultMap type="Blog" id="blogMap">
<collection property="posts" ofType="domain.blog.Post">
 <id property="id" column="post_id"/>
 <result property="subject" column="post_subject"/>
 <result property="body" column="post_body"/>
</collection>
</resultMap>
```

集合元素和关联元素几乎是一样的，它们相似的程度很高，所以没有必要再重复介绍集合元素的相似部分。

一个博客（Blog）只有一个作者（Author）。但一个博客有很多文章（Post）。在博客类中，这可以用下面的写法来表示：

```
private List<Post> posts;
```

1.2.4 动态 SQL

1. if

使用动态 SQL（即 HQL）最常见的情景是根据条件包含 where 子句的一部分。例如：

```xml
<select id="findActiveBlogWithTitleLike" resultType="Blog">
  SELECT * FROM BLOG
  WHERE state='ACTIVE'
  <if test="title!=null">
    AND title like #{title}
  </if>
</select>
```

上述语句提供了可选的查找文本功能。如果不传入"title"参数，那么所有处于"ACTIVE"状态的 BLOG 都会返回；如果传入了"title"参数，那么就会对"title"一列进行模糊查找并返回对应的 BLOG 结果。

如果希望通过"title"和"author"两个参数进行可选搜索该怎么办呢？首先，将语句名称修改为更名副其实的名称；然后，只需要加入另一个条件即可。例如：

```xml
<select id="findActiveBlogLike" resultType="Blog">
  SELECT * FROM BLOG WHERE state='ACTIVE'
  <if test="title!= null">
    AND title like #{title}
```

```xml
    </if>
    <if test="author !=null and author.name !=null">
      AND author_name like #{author.name}
    </if>
</select>
```

2. choose、when、otherwise

有时，我们不想使用所有的条件，而只是想从多个条件中选择一个条件使用。针对这种情况，MyBatis 提供了 choose 元素，它有点像 Java 中的 switch 语句。

还是上面的例子，但是策略变为：传入了"title"就按"title"查找，传入了"author"就按"author"查找的情形。若两者都没有传入，则返回标记为 featured 的 BLOG。

```xml
<select id="findActiveBlogLike" resultType="Blog">
  SELECT * FROM BLOG WHERE state='ACTIVE'
  <choose>
    <when test="title!=null">
      AND title like #{title}
    </when>
    <when test="author!=null and author.name!=null">
      AND author_name like #{author.name}
    </when>
    <otherwise>
      AND featured=1
    </otherwise>
  </choose>
</select>
```

where 元素只会在子元素返回任何内容的情况下才插入"WHERE"子句。而且，若子句的开头为"AND"或"OR"，where 元素也会将它们去除。例如：

```xml
<select id="findActiveBlogLike" resultType="Blog">
  SELECT * FROM BLOG
  <where>
    <if test="state!=null">
        state=#{state}
    </if>
    <if test="title!=null">
      AND title like #{title}
    </if>
    <if test="author!=null and author.name!=null">
      AND author_name like #{author.name}
    </if>
  </where>
</select>
```

可以通过自定义 trim 元素来定制 where 元素的功能，例如：

```xml
<trim prefix="WHERE" prefixOverrides="AND|OR">
  ...
</trim>
```

prefixOverrides 属性会忽略通过管道符分隔的文本序列。

set 元素可以用于动态包含需要更新的列，忽略其他不更新的列。例如：

```xml
<update id="updateAuthorIfNecessary">
  update Author
    <set>
      <if test="username!=null">username=#{username},</if>
      <if test="password!=null">password=#{password},</if>
      <if test="email!=null">email=#{email},</if>
      <if test="bio!=null">bio=#{bio}</if>
    </set>
  where id=#{id}
</update>
```

set 元素会动态地在行首插入 SET 关键字，并会删除额外的逗号。

3. foreach

动态 SQL 的另一个常见使用场景是对集合进行遍历（尤其是在构建 IN 条件语句时），例如：

```xml
<select id="selectPostIn" resultType="domain.blog.Post">
  SELECT *
  FROM POST P
  WHERE ID in
  <foreach item="item" index="index" collection="list"
      open="(" separator="," close=")">
        #{item}
  </foreach>
</select>
```

4. script

要在带注解的映射器接口类中使用动态 SQL，可以使用 script 元素。例如：

```java
@Update({"<script>",
    "update Author",
    "  <set>",
    "    <if test='username!=null'>username=#{username},</if>",
    "    <if test='password!=null'>password=#{password},</if>",
    "    <if test='email!=null'>email=#{email},</if>",
    "    <if test='bio!=null'>bio=#{bio}</if>",
    "  </set>",
    "where id=#{id}",
    "</script>"})
void updateAuthorValues(Author author);
```

5. bind

bind 元素允许在 OGNL（对象图导航语言）表达式以外创建一个变量，并将其绑定到当前的上下文。例如：

```xml
<select id="selectBlogsLike" resultType="Blog">
  <bind name="pattern" value="'%'+_parameter.getTitle()+'%'"/>
  SELECT * FROM BLOG
  WHERE title LIKE #{pattern}
```

```
</select>
```

1.3 Spring 框架

1.3.1 Spring IoC

IoC 也称为依赖注入（DI）。在此过程中，对象仅通过构造函数参数、工厂方法的参数、在构造或返回对象实例后设置的属性来定义其依赖项（即与它们一起使用的其他对象）。然后，容器在创建 Bean 时注入那些依赖项。从本质上讲，此过程是通过使用类的直接构造或诸如服务定位器模式之类的机制来控制其依赖关系的实例化或位置的 Bean 本身的逆过程（因此，称为控制反转）。

org.springframework.beans 和 org.springframework.context 是 Spring 框架中 IoC 容器的基础，BeanFactory 接口提供一种高级的配置机制，能够管理任何类型的对象。ApplicationContext 是 BeanFactory 的子接口，它能更容易集成 Spring 的 AOP 功能、消息资源处理、事件发布和特定的上下文，如在网站应用中的上下文 WebApplicationContext。

1.3.2 容器

1. 配置元数据

基于 XML 的配置元数据如下：

```
<?xml version="1.0" encoding="UTF-8"?>
<beans xmlns="http://www.springframework.org/schema/beans"
 xmlns:xsi="http://www.w3.org/2001/XMLSchema-instance"
 xsi:schemaLocation="http://www.springframework.org/schema/beans
 https://www.springframework.org/schema/beans/spring-beans.xsd">
 <bean id="userDao" class="com.blogs.dao.impl.UserDaoImpl"/>
 <bean id="blogDao" class="com.blogs.dao.impl.BlogDaoImpl"/>
 <bean id="userService"
   class="com.blogs.service.impl.UserServiceImpl">
   <property name="userDao" ref="userDao"/>
 </bean>
 <bean id="blogService"
   class="com.blogs.service.impl.BlogServiceImpl">
   <property name="blogDao" ref="blogDao"/>
   <!-- <constructor-arg ref="blogDao"/> -->   //构造函数的参数
 </bean>
</beans>
```

（1）该 id 属性是标识单个 Bean 定义的字符串。
（2）该 class 属性定义 Bean 的类型并使用完全限定的类名。
该 id 属性的值是指协作对象，在此示例中未显示用于引用协作对象的 XML。

2. 容器的使用

ApplicationContext 是一个维护 Bean 定义，以及相互依赖的注册表高级工厂的接口。通过使用方法 T getBean(String name, Class<T> requiredType)，可以检索 Bean 的实例。例如：

```
    ApplicationContext context=new ClassPathXmlApplicationContext
("applicationContext.xml");
    BlogService blogService=context.getBean("blogService", BlogService.class);
    blogService.insertOne(blog);
```

1.3.3 Spring Bean

1．依赖注入

依赖注入主要有两种实现方式：基于构造函数的依赖注入和基于 Setter 的依赖注入。

2．通过注解装配 Bean

1）@Required

@Required 注解适用于 Bean 属性设置器方法，例如：

```
public class SimpleMovieLister{
    private MovieFinder movieFinder;

    @Required
    public void setMovieFinder(MovieFinder movieFinder){
        this.movieFinder=movieFinder;
    }
    // ...
}
```

2）@Autowired

（1）将@Autowired 注解应用于构造函数，例如：

```
public class MovieRecommender{

    private final CustomerPreferenceDao customerPreferenceDao;

    @Autowired
    public MovieRecommender(CustomerPreferenceDao customerPreferenceDao){
        this.customerPreferenceDao=customerPreferenceDao;
    }

    // ...
}
```

（2）将@Autowired 注解应用于传统 setter 方法，例如：

```
public class SimpleMovieLister{

    private MovieFinder movieFinder;

    @Autowired
    public void setMovieFinder(MovieFinder movieFinder){
        this.movieFinder=movieFinder;
    }

    // ...
}
```

（3）将@Autowired注解应用于具有任意名称和多个参数的方法，例如：
```
public class MovieRecommender{

    private MovieCatalog movieCatalog;

    private CustomerPreferenceDao customerPreferenceDao;

    @Autowired
    public void prepare(MovieCatalog movieCatalog,
            CustomerPreferenceDao customerPreferenceDao){
        this.movieCatalog=movieCatalog;
        this.customerPreferenceDao=customerPreferenceDao;
    }

    // …
}
```
（4）将@Autowired注解应用于字段，甚至将其与构造函数混合使用，例如：
```
public class MovieRecommender{

    private final CustomerPreferenceDao customerPreferenceDao;

    @Autowired
    private MovieCatalog movieCatalog;

    @Autowired
    public MovieRecommender(CustomerPreferenceDao customerPreferenceDao){
        this.customerPreferenceDao=customerPreferenceDao;
    }

    // …
}
```
通过将@Autowired注解添加到需要该类型数组的字段或方法中，来提供ApplicationContext中所有特定类型的Bean，例如：
```
public class MovieRecommender{

    @Autowired
    private MovieCatalog[] movieCatalogs;

    // …
}
```
默认行为是将带注解的方法、构造函数和字段视为指示所需的依赖项。

3．Bean的使用域

表1-1描述了Bean的作用范围。

表 1-1

范围	描述
singleton	默认将每个 Spring IoC 容器的单个 Bean 定义范围限定为单个对象实例
prototype	将单个 Bean 定义的作用域限定为任意数量的对象实例
request	将单个 Bean 定义的范围限定为单个 HTTP 请求的生命周期。也就是说，每个 HTTP 请求都有一个在单个 Bean 定义后面创建的 Bean 实例。仅在可感知网络的 Spring ApplicationContext 中有效
session	将单个 Bean 定义的范围限定为 HTTP Session 的生命周期。仅在可感知网络的 Spring ApplicationContext 上下文中有效
application	将单个 Bean 定义的范围限定为 ServletContext 的生命周期。仅在可感知网络的 Spring ApplicationContext 上下文中有效
websocket	将单个 Bean 定义的范围限定为 WebSocket 的生命周期。仅在可感知网络的 Spring ApplicationContext 上下文中有效

1.4 SpringMVC 框架

1.4.1 SpringMVC 开发流程

SpringMVC 框架与其他很多 Web 的 MVC 框架一样：请求驱动；所有设计都围绕着一个中央 Servlet 来展开，它负责把所有请求分发到控制器；同时提供其他 Web 应用开发所需要的功能。不过 Spring 的中央处理器和 DispatcherServlet 能做的比这更多，它与 Spring IoC 容器做到了无缝集成，这意味着，Spring 提供的任何特性，在 SpringMVC 中都可以使用。

如图 1-5 所示为 SpringMVC 的 DispatcherServlet 处理请求的工作流。

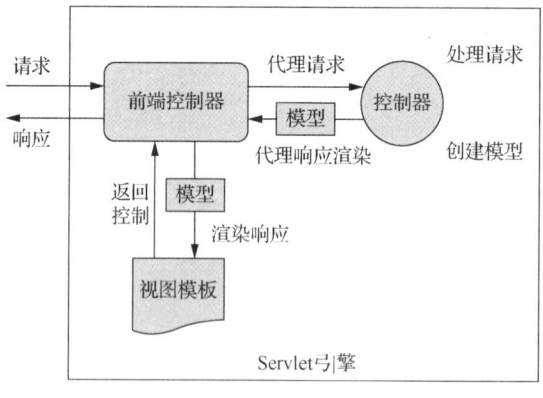

图 1-5

1.4.2 分派器

DispatcherServlet 其实就是一个 Servlet（它继承自 HttpServlet 基类），同样也需要在 Web 应用的 web.xml 配置文件下声明。需要在 web.xml 文件中将希望 DispatcherServlet 处理的请求映射到对应的 URL 上去。例如：

```
<web-app>
  <listener>
```

```xml
        <listener-class>org.springframework.web.context.ContextLoaderListener</listener-class>
    </listener>
    <context-param>
      <param-name>contextConfigLocation</param-name>
      <param-value>/WEB-INF/app-context.xml</param-value>
    </context-param>
    <servlet>
      <servlet-name>app</servlet-name>
      <servlet-class>org.springframework.web.servlet.DispatcherServlet</servlet-class>
      <init-param>
        <param-name>contextConfigLocation</param-name>
        <param-value></param-value>
      </init-param>
      <load-on-startup>1</load-on-startup>
    </servlet>
    <servlet-mapping>
      <servlet-name>app</servlet-name>
      <url-pattern>/app/*</url-pattern>
    </servlet-mapping>
</web-app>
```

在 SpringMVC 中，每个 DispatcherServlet 都持有一个自己的上下文对象 WebApplicationContext，它又继承了根（root）WebApplicationContext 对象中已经定义的所有 Bean。这些继承的 Bean 可以在具体的 Servlet 实例中被重载，在每个 Servlet 实例中也可以定义其 scope 下的新 Bean。

在 DispatcherServlet 的初始化过程中，SpringMVC 会在 Web 应用的 WEB-INF 目录下查找一个名为[servlet-name]-servlet.xml 的配置文件，并创建其中所定义的 Bean。如果在全局上下文中存在相同名称的 Bean，则它们将被新定义的同名 Bean 覆盖。

1.4.3 控制器

1. Controller 类

控制器作为应用程序逻辑的处理入口，它会负责去调用已经实现的一些服务。通常，一个控制器会接收并解析用户的请求，然后把它转换为一个模型交给视图，由视图渲染出页面最终呈现给用户。Spring 对控制器的定义非常宽松。

2. 使用@Controller 注解定义一个控制器

@Controller 注解表明了一个类是作为控制器的角色而存在的。Spring 不要求继承任何控制器基类，也不要求实现 Servlet 的那套 API。

分派器（DispatcherServlet）会扫描所有注解了@Controller 的类，检测其中通过@RequestMapping 注解配置的方法。

当然也可以不使用@Controller 注解而显式地去定义被注解的 Bean，这点通过标准的 Spring Bean 的定义方式，在 dispather 的上下文属性下配置即可做到。

Spring 支持 classpath 路径下组件类的自动检测，以及对已定义 Bean 的自动注册。需要在配置中加入组件扫描的配置代码来开启框架对注解控制器的自动检测。例如：

```xml
<?xml version="1.0" encoding="UTF-8"?>
<beans xmlns="http://www.springframework.org/schema/beans"
  xmlns:xsi="http://www.w3.org/2001/XMLSchema-instance"
  xmlns:p="http://www.springframework.org/schema/p"
  xmlns:context="http://www.springframework.org/schema/context"
  xsi:schemaLocation="
    http://www.springframework.org/schema/beans
    https://www.springframework.org/schema/beans/spring-beans.xsd
    http://www.springframework.org/schema/context
    https://www.springframework.org/schema/context/spring-context.xsd">

  <context:component-scan base-package="org.example.web"/>

  <!-- … -->

</beans>
```

3. 使用@RequestMapping 注解映射请求路径

可以使用@RequestMapping注解将请求 URL 映射到整个类上或某个特定的处理器方法上。一般来说，类级别的注解负责将一个特定（或符合某种模式）的请求路径映射到一个控制器上，同时通过方法级别的注解来细化映射，即根据特定的 HTTP 请求方法（"GET""POST"方法等）、HTTP 请求中是否携带特定参数等条件，将请求映射到匹配的方法上。例如：

```java
@Controller
@RequestMapping("/appointments")
public class AppointmentsController{
  private final AppointmentBook appointmentBook;

  @Autowired
  public AppointmentsController(AppointmentBook appointmentBook){
    this.appointmentBook=appointmentBook;
  }

  @RequestMapping(method=RequestMethod.GET)
  public Map<String, Appointment> get(){
    return appointmentBook.getAppointmentsForToday();
  }

  @RequestMapping(path="/{day}", method=RequestMethod.GET)
  public Map<String, Appointment> getForDay(@PathVariable @DateTimeFormat(iso=ISO.DATE) Date day, Model model){
    return appointmentBook.getAppointmentsForDay(day);
  }
```

```
    @RequestMapping(path="/new", method=RequestMethod.GET)
    public AppointmentForm getNewForm(){
     return new AppointmentForm();
    }

    @RequestMapping(method=RequestMethod.POST)
    public String add(@Valid AppointmentForm appointment, BindingResult result)
{
     if(result.hasErrors()){
      return "appointments/new";
     }
     appointmentBook.addAppointment(appointment);
     return "redirect:/appointments";
    }
   }
```

类级别的@RequestMapping 注解并不是必需的，若不配置，则所有的路径都是绝对路径，而非相对路径。

1.4.4 视图解析

所有 Web 应用的 MVC 框架都提供了视图相关的支持。有两个接口在 Spring 处理视图相关事宜时至关重要，分别是视图解析器接口 ViewResolver 和视图接口本身 View。视图解析器接口 ViewResolver 负责处理视图名与实际视图之间的映射关系。视图接口 View 负责准备请求，并将请求的渲染交给某种具体的视图技术实现。

1. 使用 ViewResolver 接口解析视图

若使用的是JSP视图技术,那么可以使用一个基于URL的视图解析器UrlBasedViewResolver。这个视图解析器会将 URL 解析成一个视图名，并将请求转交给请求分发器来进行视图渲染。例如：

```
<bean id="viewResolver" class="org.springframework.web.servlet.view.
UrlBasedViewResolv er">
    <property name="viewClass" value="org.springframework.web.servlet.view.
JstlView"/>
    <property name="prefix" value="/WEB-INF/jsp/"/>
    <property name="suffix" value=".jsp"/>
</bean>
```

2. 视图重定向

强制重定向的一种方法是,在控制器中创建并返回一个Spring重定向视图RedirectView 的实例。它会使 DispatcherServlet 放弃使用一般的视图解析机制，因为已经返回一个（重定向）视图给 DispatcherServlet 了，所以它会构造一个视图来满足渲染的需求。紧接着 RedirectView 会调用 HttpServletResponse.sendRedirect( )方法，发送一个 HTTP 重定向响应给客户端浏览器。

3. 向重定向目标传递数据

当前请求 URL 中的模板变量会在填充重定向 URL 时自动对应用可见，而不需要显式地在 Model 或 RedirectAttributes 中再添加属性。例如：

```java
@RequestMapping(path="/files/{path}", method=RequestMethod.POST)
public String upload(…){
  // …
  return "redirect:files/{path}";
}
```

4. 重定向前缀——redirect:

一个特别的视图名前缀能完成这个解耦，如 redirect:。如果返回的视图名中含有 redirect: 前缀，那么 UrlBasedViewResolver（及它的所有子类）就会接收到这个信号，意识到这里需要发生重定向。然后视图名剩下的部分会被解析成重定向 URL。

1.4.5 ModelAndView 类

ModelAndView 类构造方法可以指定返回的页面名称，使用 addObject( ) 设置需要返回的值。例如：

```java
public class DisplayShoppingCartController implements Controller {
  public ModelAndView handleRequest(HttpServletRequest request, HttpServletResponse response) {
    List cartItems=(List)request.getAttribute("cartItems");//拿到一个CartItem对象的列表
    User user=(User)request.getAttribute("user");//拿到当前购物的用户User
    ModelAndView mav=new ModelAndView("displayShoppingCart"); //逻辑视图名
    mav.addObject(cartItems);
    mav.addObject(user);
    return mav;
  }
}
```

第 2 章
Bootstrap 页面开发：注册页面

2.1 实验目标

（1）掌握 Bootstrap 的下载和引入。
（2）掌握 Bootstrap 基本样式的使用方法。
（3）掌握 Bootstrap 插件的使用方法。
（4）掌握 Bootstrap 组件的使用方法。
（5）综合应用 Bootstrap 框架，开发门户网站的"注册页面"。

2.2 实验任务

模拟门户网站的注册页面，页面顶部是头部导航栏，中间是 form 表单，底部是页脚。使用 Bootstrap 的导航栏组件来制作页头，使用 Bootstrap 的表单制作注册表单，注册表单内使用 Bootstrap 输入框组、按钮组件制作表单项，页面效果如图 2-1 所示。

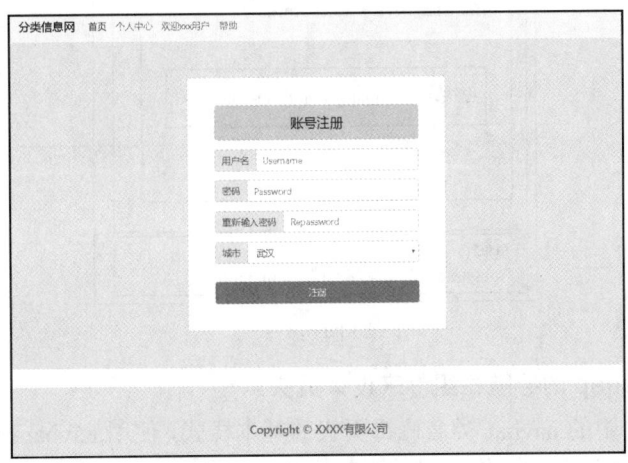

图 2-1

2.3 设计思路

使用 HBuilder 创建新项目 portal，项目文件如表 2-1 所示。

表 2-1

类型	文件	说明
html 文件	register.html	注册页 html 页面
	footer.html	公共脚部 html 页面
css 文件	css/bootstrap.min.css	Bootstrap css 文件
js 文件	js/common.js	加载脚部页面 footer.html
	js/bootstrap.min.js	Bootstrap js 文件
	js/jquery-3.2.1.min.js	jQuery 库，bootstrap 依赖

1. 创建 register.html，在页面中引入 Bootstrap

（1）在 Bootstrap 官网（https://www.bootcss.com/）下载 bootstrap.min.js 文件和 bootstrap.min.css 文件。

（2）在 jQuery 官网（https://jquery.com/）下载 jquery-3.4.1.min.js 文件。

（3）将 bootstrap.min.js 文件和 jquery-3.4.1.min.js 文件放入项目的 js 文件夹中。

（4）在 register.html 文件中引入 jquery-3.4.1.min.js 文件、bootstrap.min.js 文件和 bootstrap.min.css 文件。

2. 设计页面整体结构

（1）使用语义标签<header>、<article>、<footer>搭建页面主体结构。

（2）在<header>标签中编写页头，在<article>标签中编写注册表单，在<footer>标签中编写页脚，如图 2-2 所示。

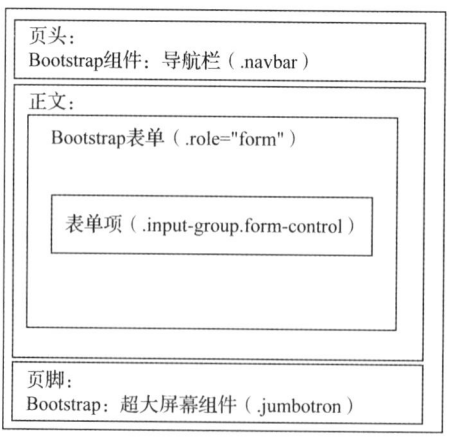

图 2-2

3. 使用 Bootstrap 的导航栏组件来设计页头

使用 Bootstrap 中的.navbar 类名修饰页头的基本样式，使用.navbar-brand 类名修饰网站标题的样式，使用.navbar-nav 修饰导航栏的样式，如图 2-3 所示。

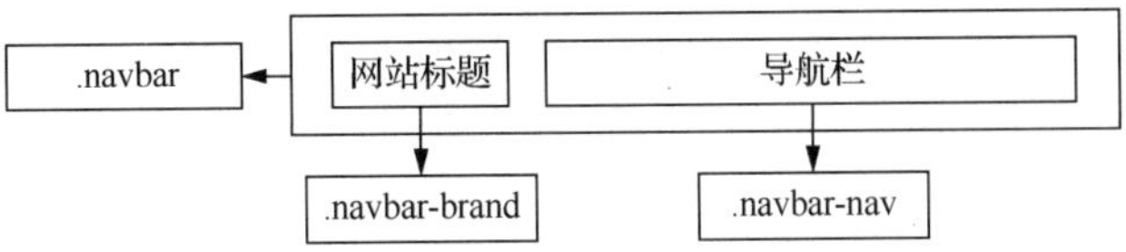

图 2-3

4．使用 Bootstrap 的表单设计注册栏

注册栏内使用输入框组、按钮组件进行设计。
（1）给正文部分<article>元素添加.jumbotron 类名，为注册栏添加更多的外边距。
（2）注册栏 form 表单盒子设计。
① 使用.card 类给盒子添加白色背景。
② 使用.col-6 类设置盒子的宽度占 6 个栅格。
（3）注册栏标题设计：使用.bg-secondary 类给标题设置一个背景色。
（4）form 表单设计。
① 向<form>元素添加 role 属性。
② 把 lable 标签和表单元素放在一个带有 class 为 input-group 的<div>中，获取最佳间距。
③ 向所有的表单元素添加 class="form-control"类来修饰文本框的样式。
④ 使用 Bootstrap 中的.btn 和.btn-success 类名修饰"注册"按钮的样式。
页面设计如图 2-4 所示。

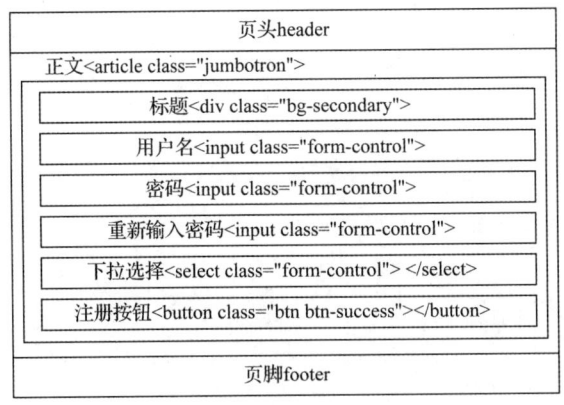

图 2-4

5．设计页脚

（1）给<footer>标签上加.jumbotron 类，给页脚添加一个背景色。
（2）使用.text-muted 类名修饰页脚的文字颜色。

2.4 实验实施（跟我做）

步骤 1：引入 Bootstrap 文件

下载 Bootstrap 资源包和 jQuery 资源包，将相应文件放入项目的 js 文件夹和 css 文件夹中。

打开 register.html 文件，通过<link>标签引入 Bootstrap 的样式文件，通过<script>标签引入 Bootstrap 的 js 文件；由于 Bootstrap 的 js 文件是基于 jQuery 的，在引入 Bootstrap 的 js 文件前先引入 jQuery 文件。

```html
<!DOCTYPE html>
<html>
<head>
<meta charset="UTF-8">
<!-- 导入 Bootstrap 样式文件 -->
<link rel="stylesheet" href="css/bootstrap.min.css">
<title>注册页面</title>
</head>
<body>
<!-- 导入 jQuery 文件 -->
<script src="js/jquery-3.4.1.min.js"></script>
<!-- 导入 Bootstrap 的 js 文件 -->
<script src="js/bootstrap.min.js"></script>
</body>
</html>
```

步骤 2：制作头部导航栏

使用 Bootstrap 的导航栏组件（.navbar）制作导航栏，代码如下：

```html
<header>
<!-- 导航栏 -->
<nav class="navbar navbar-light bg-light">
<a class="navbar-brand" href="#">分类信息网</a>
<ul class="nav">
<li class="nav-item active">
<a class="nav-link" href="#">首页</a>
</li>
<li class="nav-item">
<a class="nav-link" href="#">个人中心</a>
</li>
<li class="nav-item">
<a class="nav-link" href="#">欢迎×××用户</a>
</li>
<li class="nav-item">
<a class="nav-link" href="#">帮助</a>
</li>
</ul>
</nav>
</header>
```

页头导航栏效果如图 2-5 所示。

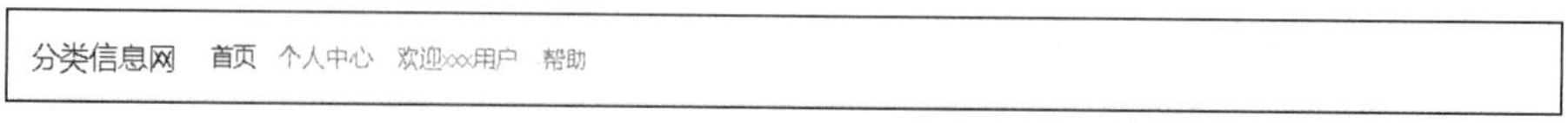

分类信息网　首页　个人中心　欢迎×××用户　帮助

图 2-5

步骤 3：制作注册栏

1．制作注册栏外围盒子

使用.col-6 类设置盒子的宽度占 6 个栅格，使用.card 类给盒子添加白色背景。

```
<article class="jumbotron">
<div class="container" align="center">
<div class="col-6 card p-5">
<!-- 注册栏内容 -->
</div>
</div>
</article>
```

2．制作注册栏标题

使用.bg-secondary 类给注册栏标题设置一个背景色。

```
<div class="bg-secondary mb-2 p-2" align="center">
<h4>账号注册</h4>
</div>
```

页面效果如图 2-6 所示。

图 2-6

3．向<form>元素添加 role 属性

（1）把标签和控件放在一个带有 class 为 input-group 的<div>中，这是获取最佳间距所必需的。

（2）向所有的文本元素<input>添加 class="form-control"类来修饰文本框的样式。

```
<form role="form">
<div class="input-group mb-3">
<div class="input-group-prepend">
<span class="input-group-text" id="basic-addon1">用户名</span>
</div>
<input type="text" class="form-control" placeholder="Username">
</div>
</form>
```

4．添加输入框组件

运用 Bootstrap 的输入框组的组件创建用户名、密码、重新输入密码 3 个输入框。

```
<form role="form">
<!-- 用户名输入框 -->
<div class="input-group mb-3">
<div class="input-group-prepend">
<span class="input-group-text" id="basic-addon1">用户名</span>
</div>
```

```html
<input type="text" class="form-control" placeholder="Username">
</div>
<!-- 密码输入框 -->
<div class="input-group mb-3">
<div class="input-group-prepend">
<span class="input-group-text" id="basic-addon2">密码</span>
</div>
<input type="password" class="form-control" placeholder="Password">
</div>
<!-- 重新输入密码输入框 -->
<div class="input-group mb-3">
<div class="input-group-prepend">
<span class="input-group-text" id="basic-addon3">重新输入密码</span>
</div>
<input type="password" class="form-control" placeholder="Repassword">
</div>
</form>
```

运行效果如图 2-7 所示。

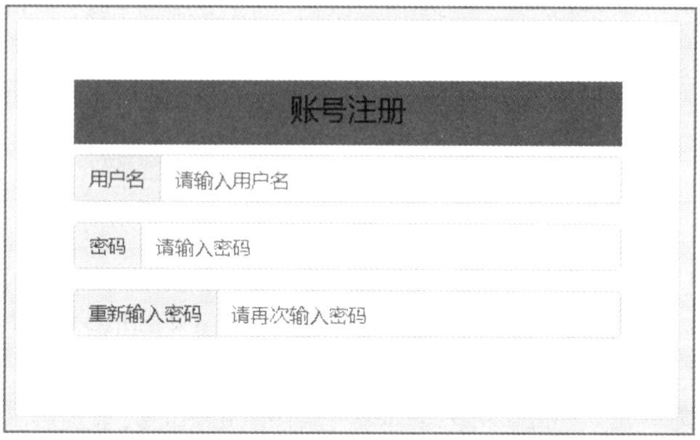

图 2-7

5．添加城市

运用 Bootstrap 的下拉选择组件创建选择城市的下拉选项。

```html
<div class="input-group mb-3">
<div class="input-group-prepend">
<span class="input-group-text" id="basic-addon3">城市</span>
</div>
<select class="form-control" name="cities" id="cities">
<option value="0">武汉</option>
<option value="1">上海</option>
<option value="2">北京</option>
</select>
</div>
```

运行效果如图 2-8 所示。

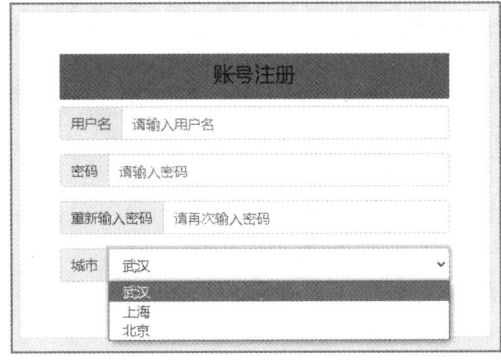

图 2-8

6．添加按钮

运用 Bootstrap 的按钮组件创建"注册"按钮，使用 Bootstrap 中的基本样式类名修饰"注册"按钮的样式。

（1）btn-success：按钮的背景色为绿色。

（2）w-100：元素宽度为 100%。

（3）mt-3：元素 margin-top 为 30px。

```
<button type="button" class="btn btn-success w-100 mt-3">注册</button>
```

运行效果如图 2-9 所示。

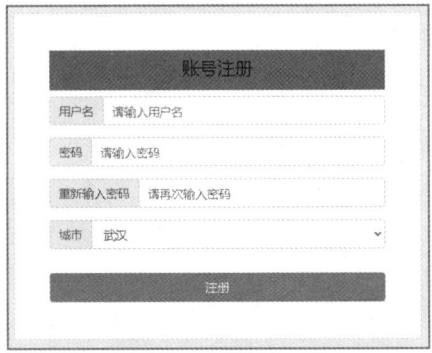

图 2-9

步骤 4：制作页脚

在 register.html 中添加 footer 元素，使用.jumbotron 类给页脚内容添加更多的外边距。

```
<footer class="jumbotron"></footer>
```

创建一个 footer.html 文件，编写页脚。

```
<h5 class="text-muted" align="center">Copyright © ××××有限公司</h5>
```

页面效果如图 2-10 所示。

图 2-10

在 js 文件夹中创建 common.js 文件，在其中编写引入 footer.html 的代码。

```
$(function(){
$("footer").load("footer.html");
});
```

在 register.html 中引入 common.js 文件，加载页脚。

```
<script src="js/common.js"></script>
```

至此，完成操作，页面运行效果如图 2-1 所示。

第 3 章 响应式网页开发：分类信息页面

3.1 实验目标

（1）掌握 Bootstrap 栅格布局的使用方法，响应移动端和 PC（personal computer，个人计算机）端。

（2）掌握 Bootstrap 基本样式的使用方法，如 .m-、.p-。

（3）掌握 Bootstrap 组件的使用方法，如导航栏组件、列表组件、媒体对象组件和分页组件。

（4）掌握 Bootstrap 插件的使用方法，如下拉插件。

（5）掌握 Bootstrap 响应式导航栏和移动端折叠导航的使用方法。

（6）综合应用 Bootstrap 框架，开发"分类信息页面"。

3.2 实验任务

分类信息页面用于显示酒类商品分类信息，页面顶部显示导航栏，中间正文部分包括订单操作栏和商品信息栏，底部页脚显示版权声明。

分类信息页面需要同时适应 PC 端和移动端，因此需要采用响应式布局。

1．PC 端布局

（1）顶部导航栏横向布局。

（2）正文订单操作栏和商品信息栏左右布局，订单操作栏在页面左侧，商品信息栏在页面右侧。

（3）订单操作栏内容从上至下纵向布局。

页面效果如图 3-1 所示。

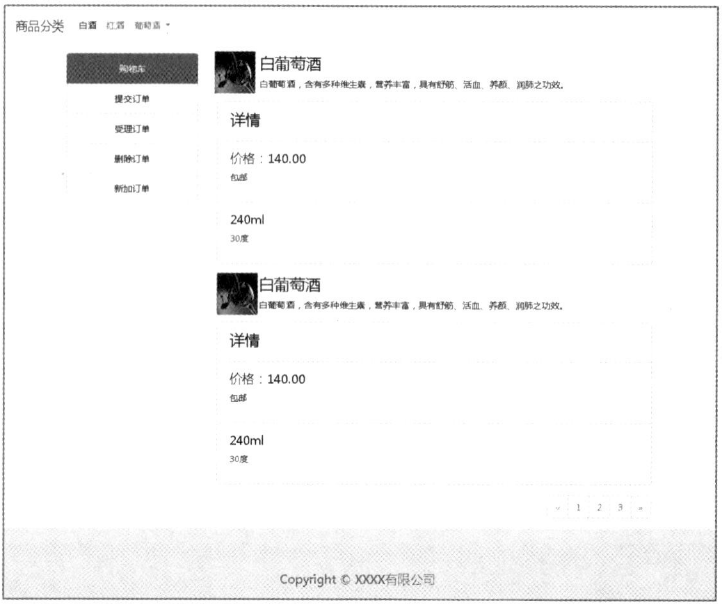

图 3-1

2．页面能够响应到移动端展示

（1）顶部导航栏横向布局。

（2）正文订单操作栏和商品信息栏上下布局，订单操作栏在页面上侧，商品信息栏在页面下侧。

（3）订单操作栏内容横向布局。

移动端页面效果如图 3-2 所示。

图 3-2

3.3 设计思路

使用 HBuilder 创建新项目 shop，项目文件如表 3-1 所示。

表 3-1

类型	文件	说明
html 文件	category.html	分类信息页 html 页面
	footer.html	公共脚部 html 页面
css 文件	css/bootstrap.min.css	Bootstrap css 文件
js 文件	js/common.js	加载脚部页面 footer.html
	js/bootstrap.min.js	Bootstrap js 文件
	js/jquery-3.2.1.min.js	jQuery 库，bootstrap 依赖

（1）创建 category.html 文件，在页面中引入 Bootstrap。
（2）使用语义化标签<header>、<aside>、<article>、<footer>搭建页面主体结构。
① <aside>标签内为订单分类列表，<article>标签内为商品信息栏。
② 在 PC 端界面，订单分类列表和商品信息栏为左右布局。在移动端界面，订单分类列表和商品信息栏为上下布局。

PC 端界面布局如图 3-3 所示。

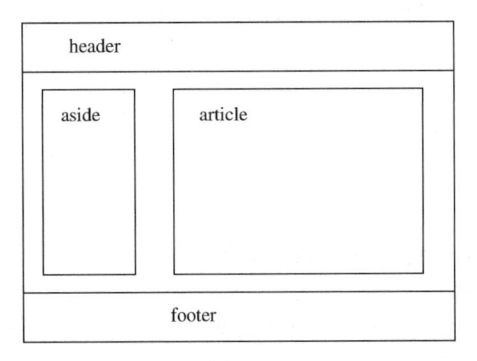

图 3-3

移动端页面布局如图 3-4 所示。

图 3-4

（3）使用 Bootstrap 的响应式导航栏组件来设计页头。

① 响应式导航栏的制作同第 2 章一样。

② 导航栏中其中一项使用下拉插件制作成下拉菜单，使用.dropdown 类名修饰下拉菜单的样式。

③ 在 PC 端界面，导航栏和网站标题在一行展示，如图 3-5 所示。

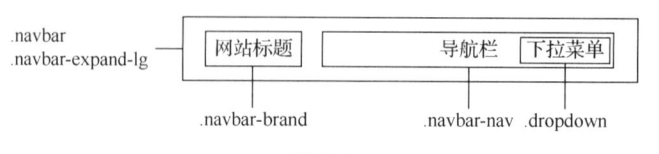

图 3-5

④ 在移动端界面，导航栏展示为折叠导航栏，如图 3-6 所示。

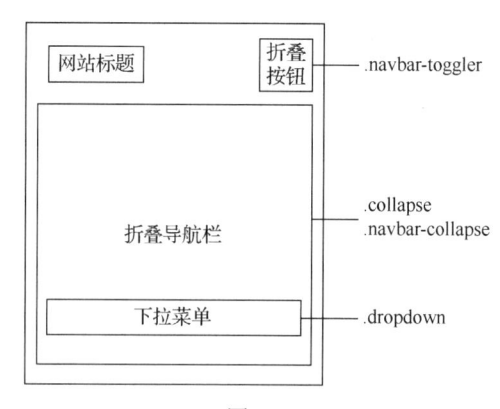

图 3-6

（4）使用 Bootstrap 栅格系统设计 PC 端的侧边栏和商品信息栏。

① 使用 Bootstrap 的栅格系统先添加类名分别为.container 和.row 的两个 div。

② .col-lg-3 类名表示当屏幕宽度≥992px 时，占栅格系统 12 列中的 3 列。

③ .col-lg-9 类名表示当屏幕宽度≥992px 时，占栅格系统 12 列中的 9 列。

PC 端的页面布局如图 3-7 所示。

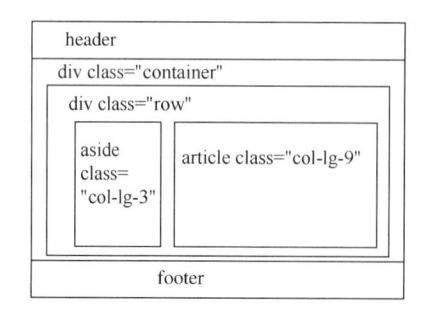

图 3-7

（5）使用 Bootstrap 的列表组件设计侧边栏的列表，如图 3-8 所示。

① 使用 Bootstrap 的.list-group 类名修饰列表组件的基本样式。

② 使用 Bootstrap 的.list-group-item 类名修饰每一项列表的样式。

```
            ┌─────────────────────────┐
            │         header          │
            │ ┌─────────────────────┐ │
            │ │列表组件              │ │
            │ │ ┌──────┐ ┌────────┐ │ │
            │ │ │.list-group-item│ │article │ │
.list-group─┤ │ ├──────┤ │        │ │ │
            │ │ │.list-group-item│ │        │ │ │
            │ │ ├──────┤ │        │ │ │
            │ │ │.list-group-item│ │        │ │ │
            │ │ └──────┘ └────────┘ │ │
            │ └─────────────────────┘ │
            │         footer          │
            └─────────────────────────┘
```

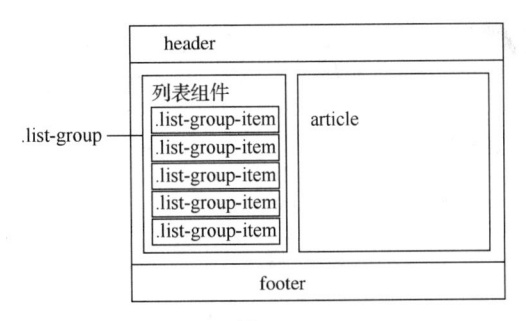

图 3-8

（6）使用 Bootstrap 的媒体对象组件、列表组件和分页组件设计商品信息栏<artilce>，如图 3-9 所示。

① 通过.media 和.media-body 类名修饰媒体对象的样式。

② 通过.list-group 和.list-group-item 类名修饰列表信息的样式。

③ 通过.pagination 和.page-item 类名修饰分页的样式。

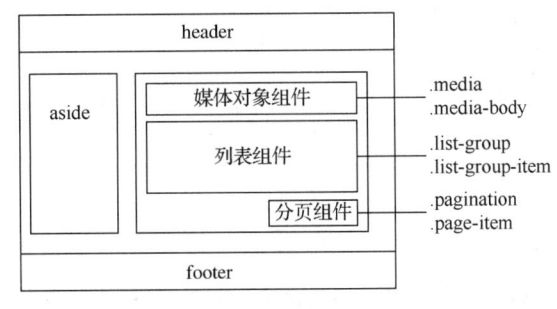

图 3-9

（7）页面响应到移动端，如图 3-10 所示。

① 使用 Bootstrap 中 display 属性类名使订单分类列表和商品信息栏上下排列，订单分类列表在一行显示。

② 使用.d-lg-block.d-none 类名显示 PC 端的订单分类列表，隐藏移动端的订单分类列表。

③ 使用.d-block.d-lg-none 类名显示移动端的订单分类列表，隐藏 PC 端的订单分类列表。

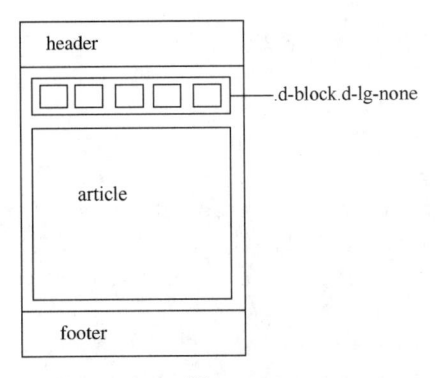

图 3-10

3.4 实验实施（跟我做）

步骤 1：引入 Bootstrap

通过<link>标签和<script>标签引入 Bootstrap。

```
<!DOCTYPE html>
<html>
<head>
<meta charset="UTF-8">
<!-- 导入 Bootstrap 样式文件 -->
<link rel="stylesheet" href=" css/bootstrap.min.css ">
<title></title>
</head>
<body>
<!-- 导入 jQuery 文件 -->
<script src="js/jquery-3.4.1.min.js"></script>
<!-- 导入 Bootstrap 的 js 文件 -->
<script src="js/bootstrap.min.js"></script>
</body>
</html>
```

步骤 2：制作页头

1. 使用 Bootstrap 的导航栏组件（.navbar）制作响应式导航栏

（1）创建响应式的导航栏（.navbar-expand-lg）。
（2）在 PC 端界面，导航栏显示在一行。
（3）在移动端界面，显示为折叠导航栏（.navbar-collapse）。代码如下：

```
<header>
<!--响应式导航栏-->
<nav class="navbar navbar-expand-lg navbar-light bg-light">
<a class="navbar-brand" href="#">商品分类</a>
<button class="navbar-toggler" type="button" data-toggle="collapse" data-target="#Nav">
<span class="navbar-toggler-icon"></span>
</button>
<!--折叠导航栏-->
<div class="collapse navbar-collapse" id="Nav">
<ul class="navbar-nav">
<li class="nav-item active"><a class="nav-link" href="#">白酒<span class="sr-only">(current)</span></a></li>
<li class="nav-item"><a class="nav-link" href="#">红酒</a></li>
<!--下拉菜单-->
<li class="nav-item dropdown">
<a class="nav-link dropdown-toggle" href="#" id="Dropdown" data-toggle="dropdown">葡萄酒</a>
```

```
<div class="dropdown-menu">
<a class="dropdown-item" href="#">白葡萄酒</a>
<a class="dropdown-item" href="#">红葡萄酒</a>
<a class="dropdown-item" href="#">紫葡萄酒</a>
</div>
</li>
</ul>
</div>
</nav>
</header>
```

PC 端页面的运行效果如图 3-11 所示。

图 3-11

移动端页面的运行效果如图 3-12 所示。

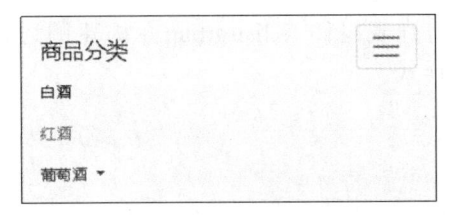

图 3-12

2．添加下拉菜单

将导航栏的葡萄酒列，使用下拉插件设置为下拉菜单（.dropdown），下拉菜单中的内容有白葡萄酒、红葡萄酒和紫葡萄酒。

```
<!--下拉菜单-->
<li class="nav-item dropdown">
<a class="nav-link dropdown-toggle" href="#" id="Dropdown" data-toggle="dropdown">葡萄酒</a>
<div class="dropdown-menu">
<a class="dropdown-item" href="#">白葡萄酒</a>
<a class="dropdown-item" href="#">红葡萄酒</a>
<a class="dropdown-item" href="#">紫葡萄酒</a>
</div>
</li>
```

添加下拉菜单后的运行效果如图 3-13 所示。

图 3-13

步骤 3：制作侧边栏和商品信息栏

（1）使用 Bootstrap 的栅格系统布局侧边栏和商品信息列表。

```html
<!-- 栅格系统最外层盒子 -->
<div class="container mt-3">
<!-- 行 -->
<div class="row">
<!-- 在 PC 端侧边栏占栅格系统 12 列的 3 列 -->
<aside class="col-lg-3">
<!-- 侧边栏的分类订单列表-->
</aside>
<!-- 在 PC 端商品信息列表占栅格系统 12 列的 9 列 -->
<article class="col-lg-9">
<!-- 商品信息列表 -->
</article>
</div>
</div>
```

（2）使用 Bootstrap 的列表组件（.list-group）构建侧边栏的分类订单，列表项使用 .list-group-item 类名设置样式。

```html
<aside class="col-lg-3">
<!--使用列表组排列-->
<div class="list-group">
<button type="button" class="list-group-item list-group-item-action active" disabled>
购物车
</button>
<button type="button" class="list-group-item list-group-item-action">提交订单</button>
<button type="button" class="list-group-item list-group-item-action">受理订单</button>
<button type="button" class="list-group-item list-group-item-action">删除订单</button>
<button type="button" class="list-group-item list-group-item-action">新加订单</button>
</div>
</aside>
```

运行效果如图 3-14 所示。

图 3-14

（3）使用 Bootstrap 的媒体对象（.media）和列表组件（.list-group）构建右侧的商品信息列表。

商品信息上部分使用 .media 媒体对象布局。

```
<!--媒体对象-->
<div class="media">
<img src="https://static.runoob.com/images/mix/img_avatar.png"
class="media-object" width="60">
<div class="media-body ml-2">
<h4 class="media-heading">白葡萄酒</h4>
<p>白葡萄酒，含有多种维生素，营养丰富，具有舒筋、活血、养颜、润肺之功效。</p>
    </div>
</div>
```

下部分"详情"使用列表组（.list-group）排列。

```
<!--列表组-->
<ul class="list-group">
<li class="list-group-item">
<h4>详情</h4>
</li>
<li class="list-group-item">
<h5>价格：140.00</h5>
<p>包邮</p>
</li>
<li class="list-group-item">
<h5>240ml</h5>
<p>30 度</p>
</li>
</ul>
```

运行效果如图 3-15 所示。

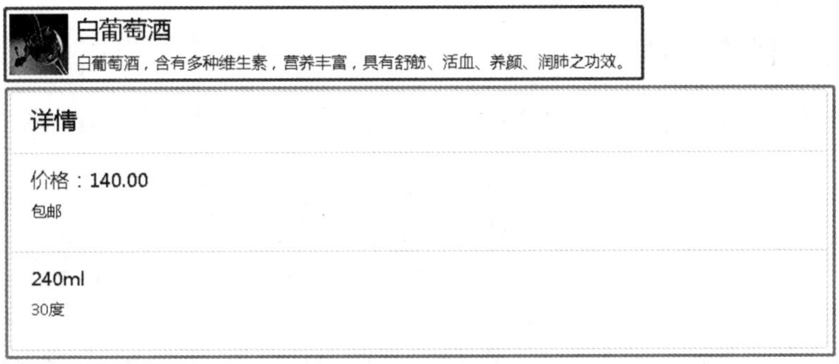

图 3-15

（4）在商品信息栏下面通过 Bootstrap 的分页组件制作商品信息的分页。

使用 .pagination 类名修饰分页框的样式，使用 .page-item 类名修饰分页页码的样式。

```
<div class="float-right mt-3">
<ul class="pagination">
<li class="page-item">
<a class="page-link" href="#">
```

```html
<span aria-hidden="true">&laquo;</span>
<span class="sr-only">Previous</span>
</a>
</li>
<li class="page-item"><a class="page-link" href="#">1</a></li>
<li class="page-item"><a class="page-link" href="#">2</a></li>
<li class="page-item"><a class="page-link" href="#">3</a></li>
<li class="page-item">
<a class="page-link" href="#">
<span aria-hidden="true">&raquo;</span>
<span class="sr-only">Next</span>
</a>
</li>
</ul>
</div>
```

运行效果如图 3-16 所示。

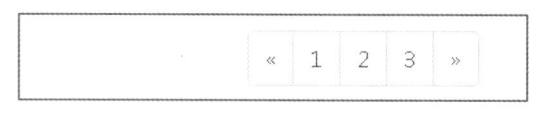

图 3-16

步骤 4：制作页脚

（1）在 category.html 中添加 footer 元素，使用.jumbotron 类给页脚内容添加更多的外边距。

```html
<footer class="jumbotron"></footer>
```

（2）在 category.html 中引入 common.js 文件，加载页脚。

```html
<script src="js/common.js"></script>
```

步骤 5：响应到移动端

（1）制作移动端横向排布的订单分类列表，使用.d-flex 对列表进行弹性布局，使列表项排列在一行。

```html
<!--移动端订单分类列表-->
<aside class="col-lg-3 d-block d-lg-none mb-2">
<div class="list-group d-flex flex-row justify-content-between align-items-center">
<button type="button" class="list-group-item active p-1" disabled>购物车</button>
<button type="button" class="list-group-item p-1">提交订单</button>
<button type="button" class="list-group-item p-1">受理订单</button>
<button type="button" class="list-group-item p-1">删除订单</button>
<button type="button" class="list-group-item p-1">新加订单</button>
</div>
</aside>
```

运行效果如图 3-17 所示。

图 3-17

（2）订单列表响应式。当在 PC 端界面时，使用.d-lg-block.d-none 类名显示 PC 端的订单分类列表，隐藏移动端的订单分类列表。

当在移动端界面时，使用.d-block.d-lg-none 类名显示移动端的订单分类列表，隐藏 PC 端的订单分类列表。

```html
<!--PC端订单分类列表-->
<aside class="col-lg-3 d-lg-block d-none">
<div class="list-group">
<button type="button" class="list-group-item active" disabled>购物车</button>
<button type="button" class="list-group-item">提交订单</button>
<button type="button" class="list-group-item">受理订单</button>
<button type="button" class="list-group-item">删除订单</button>
<button type="button" class="list-group-item">新加订单</button>
</div>
</aside>
<!--移动端订单分类列表-->
<aside class="col-lg-3 d-block d-lg-none mb-2">
<div class="list-group d-flex flex-row justify-content-between align-items-center">
<button type="button" class="list-group-item active p-1" disabled>购物车</button>
<button type="button" class="list-group-item p-1">提交订单</button>
<button type="button" class="list-group-item p-1">受理订单</button>
<button type="button" class="list-group-item p-1">删除订单</button>
<button type="button" class="list-group-item p-1">新加订单</button>
</div>
</aside>
```

运行效果如图 3-18 所示。

图 3-18

至此，完成操作，PC 端界面的运行效果如图 3-1 所示，移动端界面的运行效果如图 3-2 所示。

第 4 章 创建第一个 Java 程序

4.1 实验目标

（1）掌握 Java 开发环境的搭建。
（2）掌握 Eclipse 的安装和配置方法。
（3）掌握 Tomcat 服务器的安装和配置方法。
（4）掌握 Java 动态网站工程的创建方法。
（5）掌握 Java 动态网站工程的发布与运行方法。

4.2 实验任务

（1）下载并安装 JDK。
（2）下载并安装 Eclipse jee 集成开发工具。
（3）下载并安装 XAMPP。
（4）在 Eclipse 中配置 JDK 和 Tomcat 服务器。
（5）创建第一个 Java 动态网站工程，并使用脚本输出"Hello World"。
（6）在 Tomcat 服务器中运行工程。

4.3 设计思路

1. 工程设计

创建 Dynamic Web Project 项目——HelloWeb。

2. 文件设计

文件设计如表 4-1 所示。

表 4-1

类型	文件	说明
JSP 文件	WebContent/index.jsp	主页
配置文件	WebContent/WEB-INF/web.xml	Web 配置文件

4.4 实验实施（跟我做）

步骤 1：环境搭建

1. 下载和安装 JDK8u202

进入 JDK 官网 https://www.oracle.com/java/technologies/oracle-java-archive-downloads.html，如图 4-1 所示。

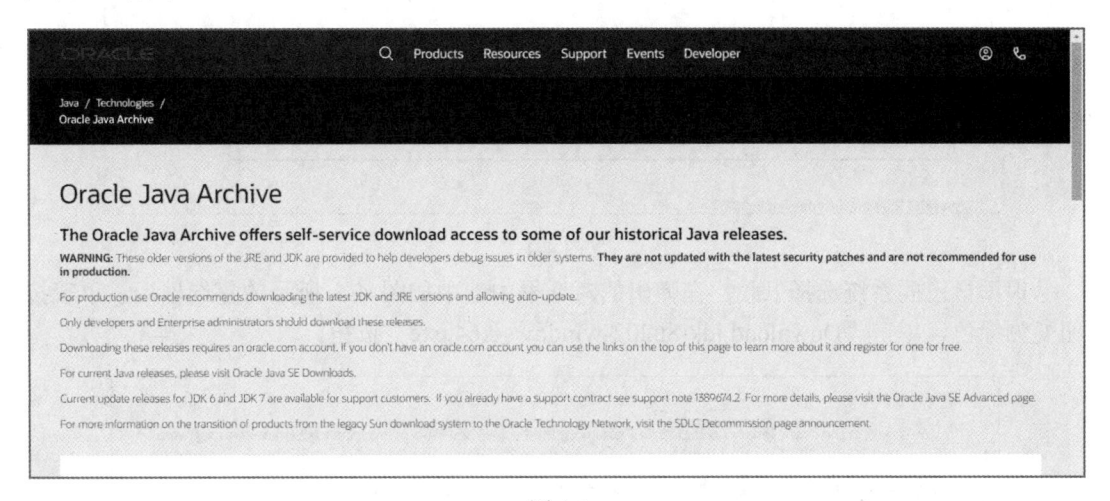

图 4-1

下滑页面找到"Java SE"，如图 4-2 所示。

图 4-2

可以看到 Java SE 8 分为两个版本，如图 4-3 所示，8u202 是最后一个 Java 开源版本。

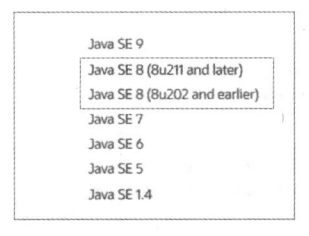

图 4-3

单击"Java SE 8(8u202 and earlier)"链接进入下载页面，如图 4-4 所示。

图 4-4

根据自己的系统选择下载，在弹出的对话框中选中如图 4-5 所示的复选框后，下载按钮变为绿色，单击"Download jdk-8u202-windows-x64.exe"按钮。

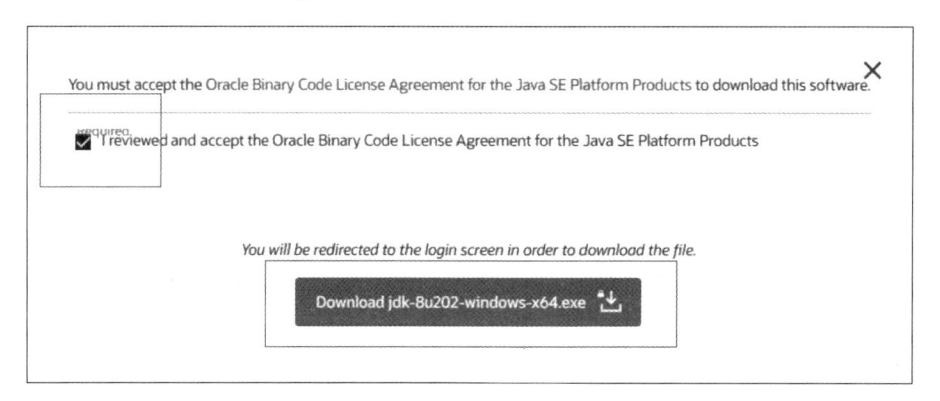

图 4-5

进入登录页面，登录系统后即可自动下载。下载完成后，双击.exe 文件默认安装。安装成功后，找到安装目录，如图 4-6 所示。

图 4-6

2. 配置 Java 环境变量

打开环境变量，添加"JAVA_HOME"系统变量，变量值为 JDK 的安装目录，如图 4-7 所示。

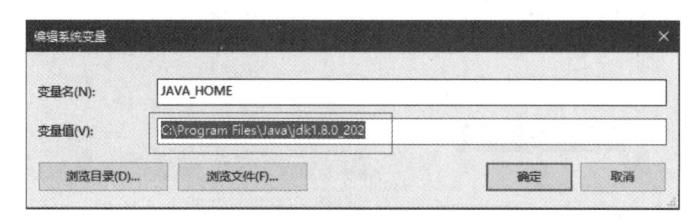

图 4-7

在环境变量中添加"%JAVA_HOME%\bin"，如图 4-8 所示。

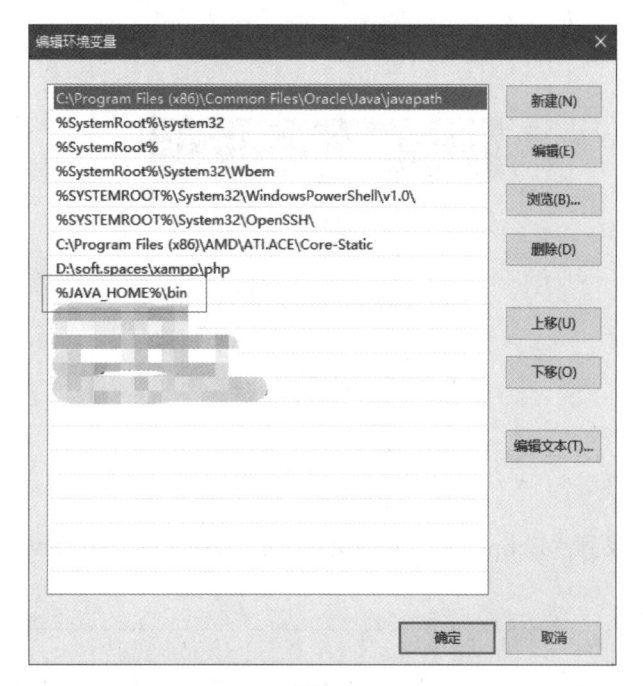

图 4-8

启动命令行，输入命令"java -version"，若显示 Java 版本信息，则表示已完成配置，如图 4-9 所示。

图 4-9

3. 下载和安装 Eclipse

进入 Eclipse 官网：https://www.eclipse.org/downloads/，如图 4-10 所示。

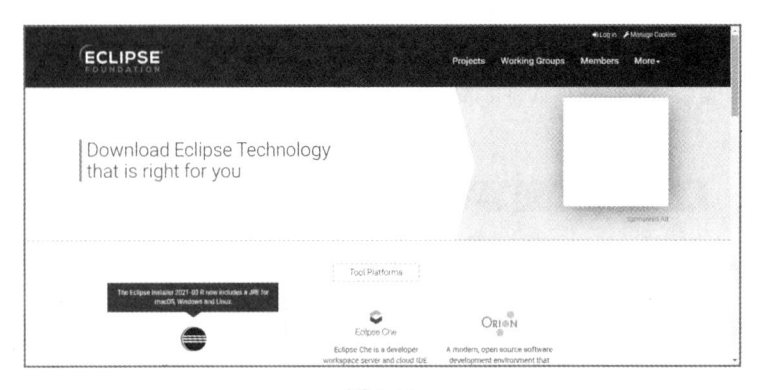

图 4-10

在 Eclipse IDE 下载按钮下找到"Download Packages"链接,如图 4-11 所示,单击该链接进入下载安装包页面。

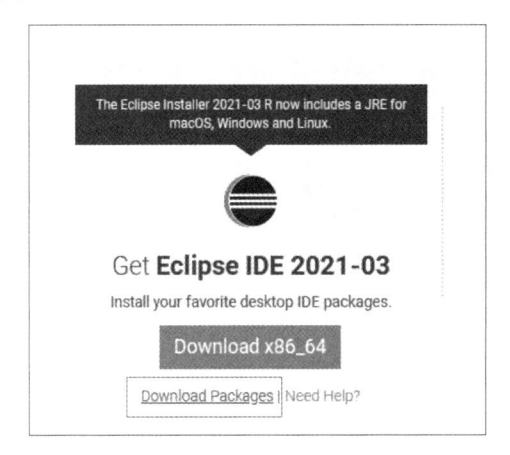

图 4-11

在工具列表中找到"Eclipse IDE for Enterprise Java and Web Developers",如图 4-12 所示。

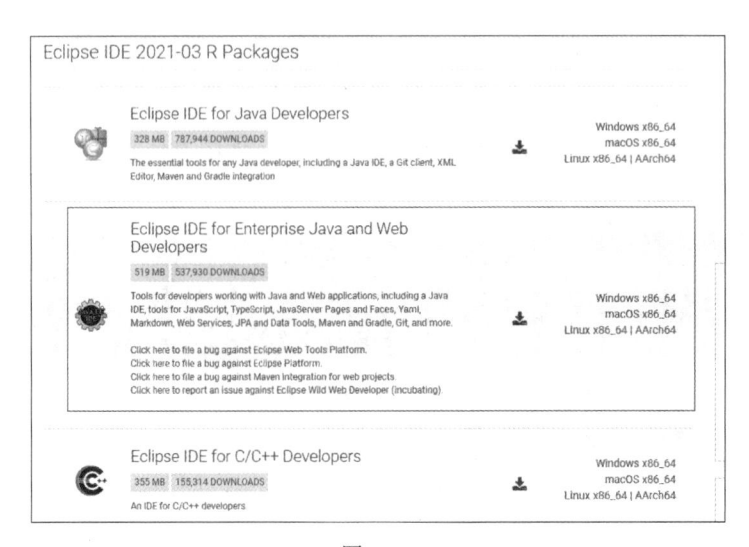

图 4-12

根据系统选择安装包下载，然后单击"Download"按钮下载即可，如图 4-13 所示。下载的安装包如图 4-14 所示。

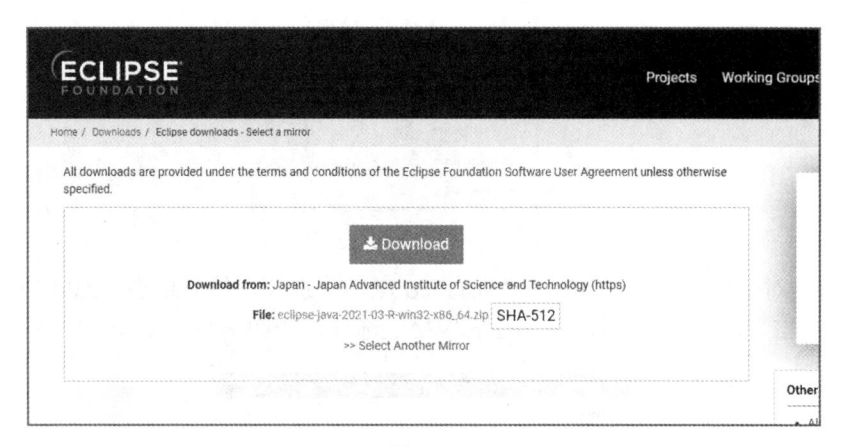

图 4-13

图 4-14

将下载的 zip 文件，直接解压到 C 盘根目录（可自行调整），解压后是"eclipse"文件夹，如图 4-15 所示。

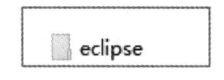

图 4-15

打开文件夹，单击"eclipse.exe"文件，即可启动 Eclipse，如图 4-16 所示。

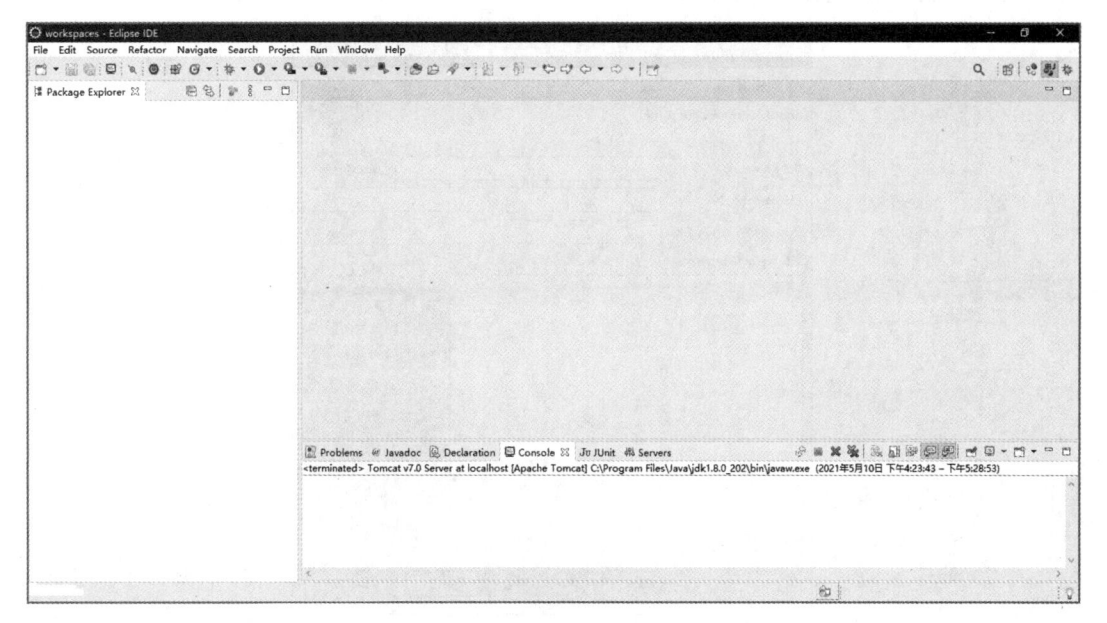

图 4-16

打开 Eclipse，选择"Window"→"Preferences"选项，如图 4-17 所示。

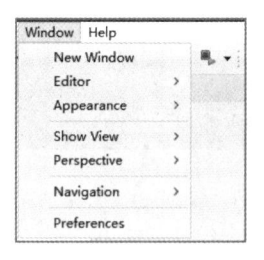

图 4-17

在弹出的对话框选择"Java"→"Installed JREs"选项，然后单击"Add"按钮，在弹出的"Add JRE"对话框中选择"Standard VM"选项，如图 4-18 所示。

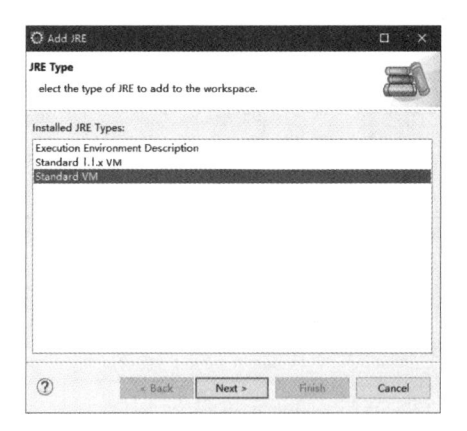

图 4-18

单击"Next"按钮，在打开的页面中单击"Directory"按钮，如图 4-19 所示，然后在弹出的对话框中选择 JDK 的根目录。

图 4-19

在弹出的"Preferences"对话框中完成 JDK 的配置，如图 4-20 所示，然后单击"Apply and Close"按钮。

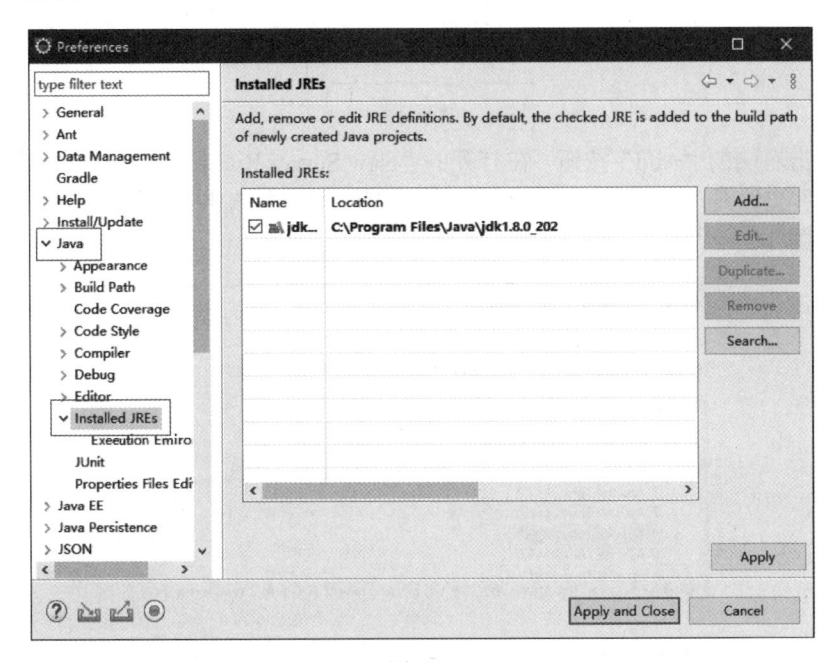

图 4-20

4．安装和配置 Tomcat

下载 Tomcat 安装程序，默认安装即可，安装成功后到安装目录中找到"tomcat"文件夹，如图 4-21 所示，表示 Tomcat 安装成功。

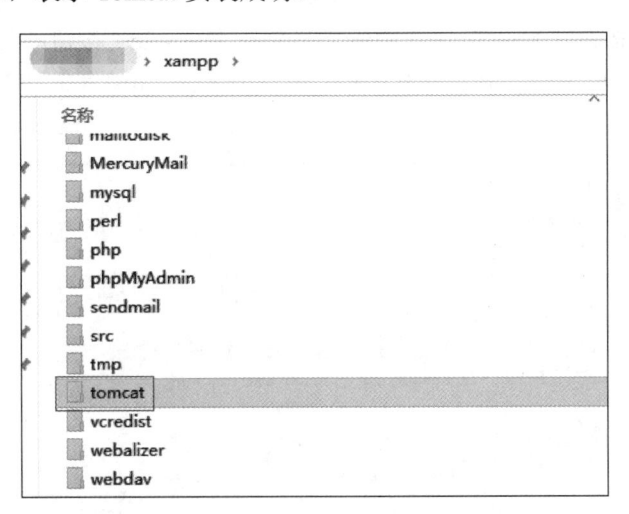

图 4-21

打开 Eclipse，选择"Window"→"Preferences"选项，在弹出的对话框左侧选择"Server"→"Runtime Environments"选项，如图 4-22 所示。

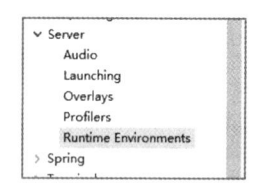

图 4-22

然后单击右侧的"Add"按钮，在打开的"New Server Runtime Environment"窗口中找到"Apache"文件夹，选择"Apache Tomcat v8.5"选项，并选中"Create a new local server"复选框，如图 4-23 所示。

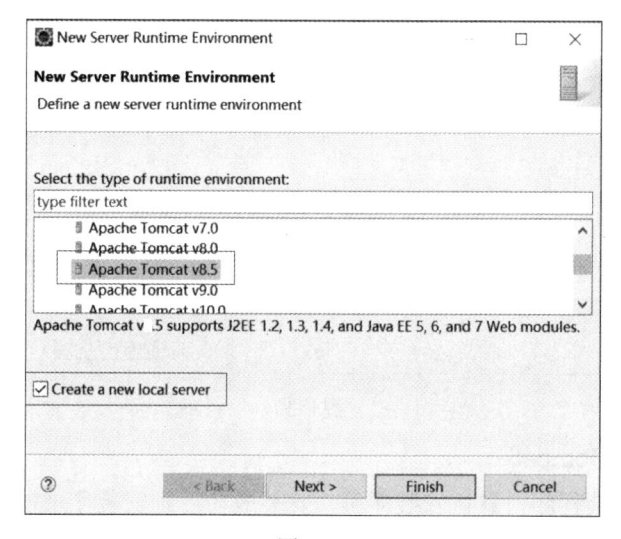

图 4-23

单击"Next"按钮，配置 Tomcat 服务器，单击"Browse"按钮，在弹出的对话框中选择 xampp 下的 tomcat 目录，在"JRE"下拉列表中选择"jre1.8.0_202"选项，如图 4-24 所示。

图 4-24

单击"Finish"按钮完成 Tomcat 服务器的配置，在右下角的"Servers"选项卡中会显示刚配置的 Tomcat 图标，如图 4-25 所示。

图 4-25

在左侧的"Package Explorer"视窗中会多出一个"Servers"目录，如图 4-26 所示。

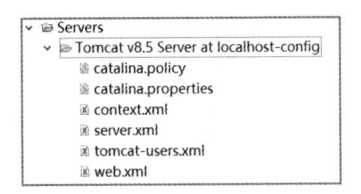

图 4-26

步骤 2：创建工程

启动 Eclipse，选择"File"→"New"→"Dynamic Web Project"工程。若在弹出的子菜单中未找到该工程，则选择"Other"选项，在弹出的"Wizards"选择向导窗口列表框中的"Web"文件夹中找到"Dynamic Web Project"，如图 4-27 所示。

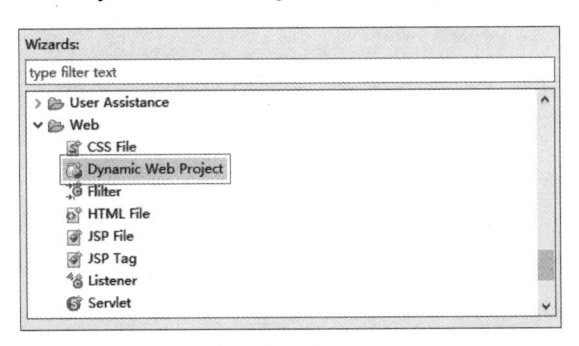

图 4-27

弹出工程创建的向导对话框，在"Project name"文本框中输入"HelloWeb"工程名，在"Target runtime"下拉列表中选择"Apache Tomcat v7.0"选项，如图 4-28 所示。

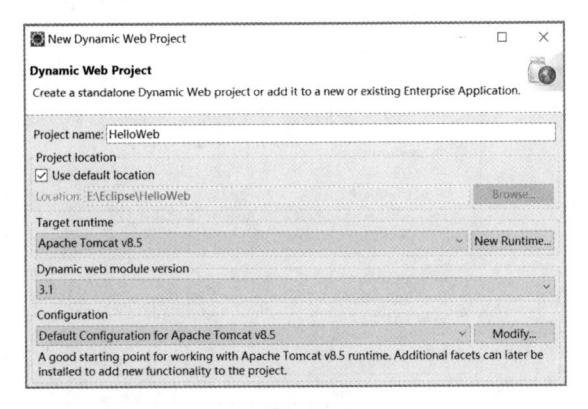

图 4-28

单击"Next"按钮，弹出如图 4-29 所示的对话框。

单击"Next"按钮，在弹出的对话框中选中"Generate web.xml deployment descriptor"复选框，向导将自动创建 web.xml 文件，如图 4-30 所示。

图 4-29

图 4-30

单击"Finish"按钮，完成创建工程的操作，在 Package Explorer 中可以看到新创建的"HelloWeb"工程，如图 4-31 所示。

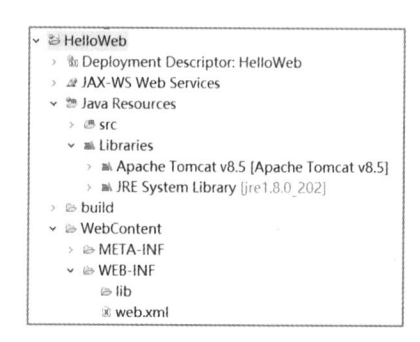

图 4-31

步骤 3：编写第一个 JSP 页面

1. 创建 index.jsp 文件

在 Package Explorer 中选择"HelloWeb"工程下的"WebContent"文件夹图标，右击，在弹出的快捷菜单中选择"New"→"JSP File"选项，若没有该选项，则选择"Other"选项，如图 4-32 所示。

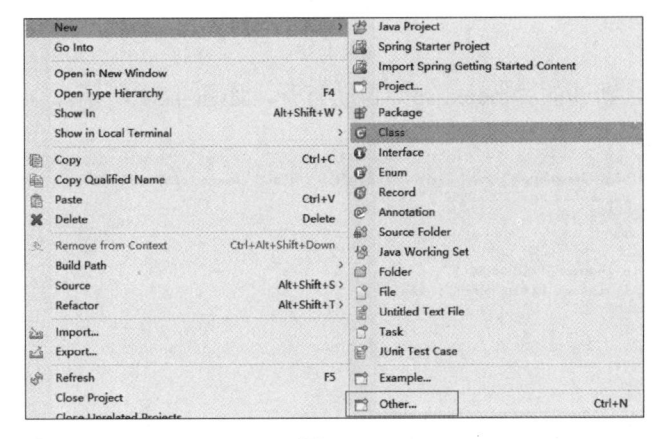

图 4-32

在弹出的"Wizards"选择向导窗口中的"Web"文件夹下找到"JSP File"，如图 4-33 所示。

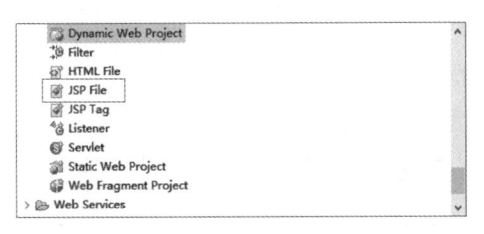

图 4-33

单击"Next"按钮，在弹出的对话框中的"File name"文本框中输入"index.jsp"，如图 4-34 所示。

图 4-34

单击"Finish"按钮，在打开的界面左侧"Package Explorer"视图中的"WebContent"文件夹下新增了"index.jsp"文件图标，如图 4-35 所示。

图 4-35

在右侧编辑区，新建的"index.jsp"自动打开，如图 4-36 所示。

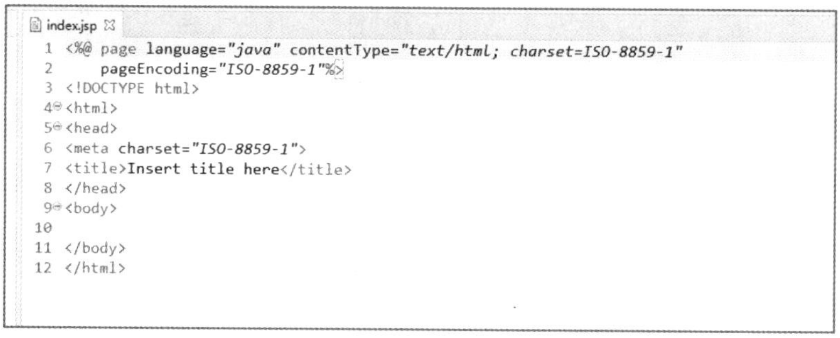

图 4-36

2．编辑代码

编写 index.jsp 页面代码，在<body></body>中添加代码"<%="Hello World" %>"，向页面输出"Hello World"。

```
<%@ page language="java" contentType="text/html; charset=ISO-8859-1"
    pageEncoding="ISO-8859-1"%>
<!DOCTYPE html>
<html>
<head>
<meta charset="ISO-8859-1">
<title>Insert title here</title>
</head>
<body>
<%="Hello World"%>
</body>
</html>
```

步骤 4：运行程序

在"Package Explorer"视窗中选择"HelloWeb"项目图标右击，在弹出的快捷菜单中选择"Run As"→"Run on Server"选项，如图 4-37 所示。

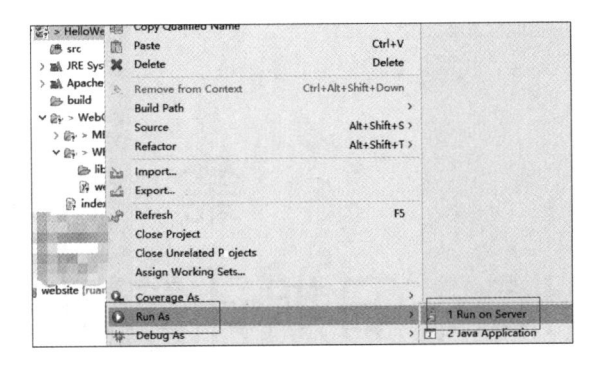

图 4-37

在打开的"Run On Server"窗口中选中"Always use this server when running this project"复选框，如图 4-38 所示。

图 4-38

单击"Finish"按钮，等待服务器启动，启动成功后在代码编辑区弹出网页，如图 4-39 所示。

图 4-39

第 5 章 类和对象：日期计算器

5.1 实验目标

（1）掌握 Java 的基础语法、编码规范。
（2）掌握 Java 的数据类型和数据类型转换的方法。
（3）掌握 Java 常量、变量的定义和使用方法。
（4）掌握 Java 基本运算符的使用方法。
（5）掌握 Java 条件判断语句、循环控制语句的使用方法。
（6）掌握 Java 数组的定义和使用方法。
（7）掌握 Java 的类和对象。
（8）掌握 JSP/Servlet 编程的方法。
（9）综合应用 Java 面向对象编程，编写"日期计算器"程序。

5.2 实验任务

使用 Java 面向对象编程，以 JSP 页面实现日期计算功能。
（1）在网页上布局输入框和"计算"按钮。
（2）根据输入的年、月、日，判断该年是平年还是闰年。
（3）计算从该年元旦到输入日一共是多少天。
（4）分别输出判断和计算结果。

例如，当在界面中输入年为 2019、月为 8 和日为 2 时，如图 5-1 所示，此时计算从"2019-01-01"到"2019-08-02"一共经过 214 天，计算结果如图 5-2 所示。

图 5-1

> 2019-08-02是平年。
> 从2019-01-01到2019-08-02共经过214天。

图 5-2

5.3 设计思路

1. 工程设计

创建 Dynamic Web Project 项目——Calculator。

2. 文件设计

创建 index.jsp 为主页面,显示日期计算器。在 index.jsp 文件中布局"日期计算器"网页,添加 3 个 <input> 输入框(年 year、月 month、日 day),添加 <button> "计算"按钮,如图 5-3 所示。添加表单 form,设置 action 属性为请求 Calculate 文件,method 为 POST。

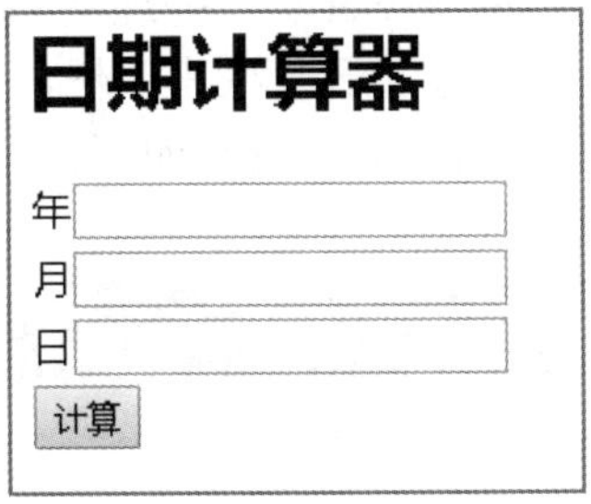

图 5-3

3. 编写日期处理逻辑

使用 com.ca.model.Calculator 类编写日期处理逻辑。

(1) 定义 Calculator 类,添加年、月、日 3 个属性,用来存储日期信息。在构造函数中初始化年、月、日的值。

(2) 添加 public boolean isLeap( ) 函数,判断日期是否为闰年。

闰年规则:能被 4 整除而不能被 100 整除,或能被 100 整除也能被 400 整除。

如果是闰年,则返回 true,否则返回 false。

(3) 添加 public boolean checkMD( ) 函数,检查日期的有效性。

使用 if 条件语句验证输入的日期是否存在。

创建数组 int[ ] month_days,将 1~12 月每月的天数存入数组对应下标的元素中,2 月默认存入 28。

调用 isLeap( ) 函数验证是否为闰年,如果为闰年则将 month_days[2] 的值设为 29。

(4) 添加 public int countTotalDay( ) 函数,计算天数。

使用 for 循环,计算从一月到输入月份上一个月的天数总和。

使用 switch 语句,判断各个月份的天数。

月份之和加上输入的天数,即为从该年元旦到这一天一共经过的天数。

日期的处理逻辑如图 5-4 所示。

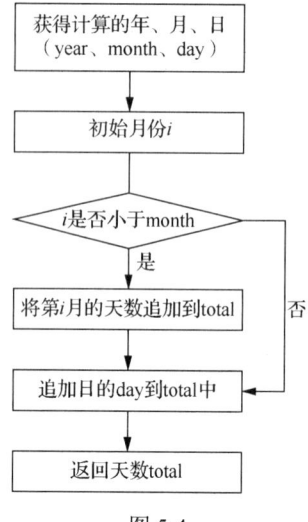

图 5-4

（5）使用 com.ca.servlet.CalculateServlet 类。处理从 POST 表单中获取用户输入的年、月、日，并创建 Calculator 类对象，检查日期的有效性，判断平年、闰年，以及该天是当年的第几天，并将结果输出。请求路由为"/Calculate"。

5.4 实验实施（跟我做）

步骤 1：创建工程

启动 Eclipse，选择"File"→"New"→"Dynamic Web Project"工程，在打开的向导窗口中的"Project name"文本框中输入"Calculator"工程名，在"Target runtime"下拉列表中选择"Apache Tomcat v8.5"选项。

单击"Next"按钮，在打开的"Web Module"配置窗口中选中"Generate web.xml deployment descriptor"复选框，向导将自动创建 web.xml 文件。

单击"Finish"按钮，完成创建工程的操作，在 Package Explorer 中可以看到新创建的 Calculator 工程。

步骤 2：编写主页

（1）使用输入框<input type="text"/>获得用户输入的年、月、日。
（2）添加"计算"按钮<input type="submit"/>。
（3）添加 form 表单，提交到请求处理 Calculate，方法为 post。

代码如下：

```
<%@ page language="java" contentType="text/html; charset=UTF-8" pageEncoding="UTF-8"%>
<html>
<head>
<meta charset="utf-8"/>
<title>日期计算器</title>
```

```html
</head>
<body>
  <header>
    <h1>日期计算器</h1>
  </header>
  <article>
    <form action="Calculate" method="post">
      <table>
        <tr>
          <td><label>年</label><input name="year" type="text"/></td>
        </tr>
        <tr>
          <td><label>月</label><input name="month" type="text"/></td>
        </tr>
        <tr>
          <td><label>日</label><input name="day" type="text"/></td>
        </tr>
        <tr>
          <td><input name="calculate" type="submit" value="计算"/></td>
        </tr>
      </table>
    </form>
  </article>
</body>
</html>
```

步骤 3：创建 Calculator 类

（1）创建 com.ca.model 包，创建 Calculator 类。
（2）添加 3 个属性。
（3）创建带参的构造方法。

```java
package com.ca.model;

public class Calculator{
  private int year;
  private int month;
  private int day;
}
```

步骤 4：判断是否为闰年

（1）在 Calculator 类中创建 isLeap( )函数，判断某年是否为闰年。
（2）公历闰年算法如下。
能被 4 整除而不能被 100 整除。
能被 100 整除也能被 400 整除。
代码如下：

```
public boolean isLeap(){
  if((this.year%4==0&&this.year%100!=0)||(this.year%100==0&&this.year%400==0)){
    return true;
  } else{
    return false;
  }
}
```

步骤 5：验证日期

（1）在 Calculator 类中创建 checkMD( )函数，编写日期计算的相关函数。

（2）调用 checkMD( )函数，验证日期。

使用 if 条件语句，判断输入的 this.month 的值是否为 1~12，this.day 的值是否为 1~31。若超过了，则返回 false。

定义 int[ ]month_days 数组，定义每月的总天数，下标 0 不使用设置为 0，使用下标 1~12 存储相应月份的天数，下标为 2 的值设为 28。

调用 isLeap( )函数，判断当前年份是否为闰年，如果为闰年，则将 month_days[2]的值设为 29。

调用 if 条件语句，判断天数是否超过对应月份的总天数，如果超过则返回 false，否则返回 true。

```
public boolean checkMD(){
  //验证月份是否为 1~12,日期是否为 1~31
  if(this.month<1||this.month>12||this.day>31||this.day<1){
    return false;
  }
  //定义数组,存放每个月的天数,数组索引号与月份对应,下标为 0 的设置为 0 不使用
  int[]month_days={0,31,28,31,30,31,30,31,31,30,31,30,31};

  //判断当前年份是否为闰年,如果为闰年,则将$days 数组下标为 2 的值设为 29
  if(this.isLeap()){
    month_days[2]=29;
  }

  //判断天数是否超过了当月的最大天数
  if(this.day>month_days[this.month]){
    return false;
  }
  return true;
}
```

步骤 6：计算天数

（1）在 Calculator 类中创建 countTotalDay( )函数，用来计算日期总天数。

（2）使用 for 循环计算从该年元旦到这一天一共经过了多少天。

```
public int countTotalDay(){
  int total=0;
  for(int i=1; i<this.month; i++){
    switch (i){
    ……
    }
  }
}
```

（3）使用 switch 语句判断每个月的总天数。
```
switch(i){
  case 1:
  case 3:
  case 5:
  case 7:
  case 8:
  case 10:
  case 12:
    month_day=31;
    break;
  case 4:
  case 6:
  case 9:
  case 11:
    month_day=30;
    break;
  case 2:
    if(isLeap()){
      month_day=29;
    } else {
      month_day=28;
    }
    break;
  default:
    break;
}
```

（4）使用 for 语句计算输入月份之前的所有月份天数之和，然后加上输入的天数，结果即为从该年元旦到输入的日期这一天一共经过的天数。
```
public int countTotalDay(){
  int total=0;
  for(int i=1; i<this.month; i++){
    int month_day=0;
    switch (i){
    ……
    }
    total+=month_day;
  }
```

```
            total=total+this.day;
            return total;
     }
```

步骤 7：处理请求

（1）创建 com.ca.servlet 包，创建 Servlet 类 CalculateServlet，计算日期天数，并输出结果。

（2）编写 doPost( )函数，从 request 中获取年、月、日，创建 Calculator 类对象。

（3）调用 checkMD( )函数判断日期天数是否正确，调用 isLeap( )函数判断当前年份是否为闰年。

（4）调用 countTotalDay( )函数，计算当前日期是这一年的第几天。

```java
package com.ca.servlet;

import java.io.IOException;
import java.io.PrintWriter;

import javax.servlet.ServletException;
import javax.servlet.annotation.WebServlet;
import javax.servlet.http.HttpServlet;
import javax.servlet.http.HttpServletRequest;
import javax.servlet.http.HttpServletResponse;

import com.ca.model.Calculator;

/**
 * Servlet implementation class CalculateServlet
 */
@WebServlet("/Calculate")
public class CalculateServlet extends HttpServlet{
  private static final long serialVersionUID=1L;

  /**
   * @see HttpServlet#HttpServlet()
   */
  public CalculateServlet(){
     super();
     // TODO Auto-generated constructor stub
  }

  /**
   * @see HttpServlet#doPost(HttpServletRequest request, HttpServletResponse
   *      response)
   */
  protected void doPost(HttpServletRequest request, HttpServletResponse response)
```

```java
        throws ServletException, IOException{
    int year=Integer.parseInt(request.getParameter("year"));
    int month=Integer.parseInt(request.getParameter("month"));
    int day=Integer.parseInt(request.getParameter("day"));
    String date=String.format("%d-%d-%d", year, month, day);
    
    response.setCharacterEncoding("UTF-8");
    response.setContentType("text/html; charset=utf-8");
    PrintWriter writer=response.getWriter();
    
    Calculator ca=new Calculator(year, month, day);
    if(ca.checkMD()){
      if(ca.isLeap()){
        writer.append(date+"是闰年。<br/>");
      } else {
        writer.append(date+"是平年。<br/>");
      }
      int total=ca.countTotalDay();
      writer.append("从"+year+"-1-1 到"+date+"共经过"+total+"天。<br/>");
    } else{
      writer.append("日期不存在。<br/>");
    }
  }
}
```

步骤 8：配置欢迎页面

打开 "WebContent/WEB-INF" 文件夹下的 web.xml 文件。修改代码，将 welcome-file 修改为 index.jsp，这样当访问项目根目录 "localhost:8080/Calculator/" 时，默认显示 index.jsp 页面。

```xml
<?xml version="1.0" encoding="UTF-8"?>
<web-app xmlns:xsi="http://www.w3.org/2001/XMLSchema-instance" xmlns="http://java.sun.com/xml/ns/javaee" xsi:schemaLocation="http://java.sun.com/xml/ns/javaee http://java.sun.com/xml/ns/javaee/web-app_3_0.xsd" id="WebApp_ID" version="3.0">
    <display-name>Calculator</display-name>
    <welcome-file-list>
      <welcome-file>index.jsp</welcome-file>
    </welcome-file-list>
</web-app>
```

步骤 9：运行效果

（1）运行 Calculator 程序，URL 为 http://localhost:8080/Calculator/，显示主页，如图 5-5 所示。

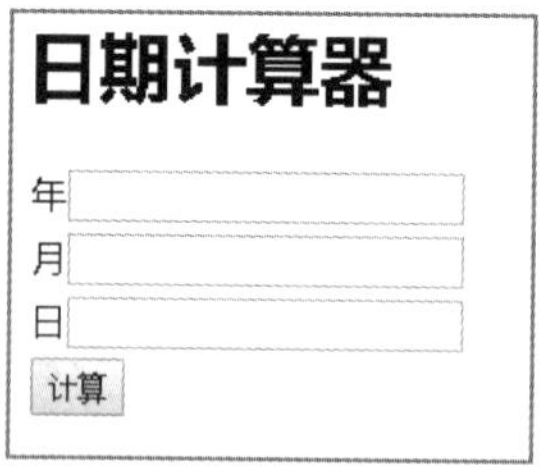

图 5-5

（2）在"年""月""日"输入框分别输入 2019、08、02，如图 5-6 所示，单击"计算"按钮，输出结果，如图图 5-7 所示。

图 5-6

图 5-7

错误日期的输出结果如图 5-8 和图 5-9 所示。

图 5-8

图 5-9

第6章 Java Web 编程（JSP）：在线投票

6.1 实验目标

（1）掌握 Servlet 类的创建方法，以及 doPost 方法的定义。
（2）掌握@WebServlet 注解的使用方法。
（3）掌握 Request 和 Response 对象的使用方法。
（4）掌握 Servlet 中请求转发的实现方法。
（5）了解 Request 对象编码格式的设置。
（6）掌握 JSP 页面 Java 类的导入方法、Java 代码的嵌入方法、Java 变量的显示方法。
（7）了解 JSP/Servlet 中文乱码的处理方法。
（8）了解 Java 时间格式化类的使用方法。
（9）综合应用 Java 面向对象编程，编写"在线投票"程序。

6.2 实验任务

编写一个在线投票程序，用户进入投票页面，输入自己的姓名并选择认为可获得世界杯冠军的国家，然后提交。

程序包含姓名输入页面、投票页面和结果页面。

（1）姓名输入页面：页头显示"输入用户名"，页面内容显示一个输入框和"下一步"按钮，页脚显示当前时间，如图 6-1 所示。

图 6-1

(2)投票页面:页头显示"谁能获得世界杯冠军?",页面内容使用单选按钮列出可能获得世界杯冠军的国家(中国、美国、巴西、意大利、英国、德国、西班牙和法国),并显示一个"确定"按钮,如图 6-2 所示。

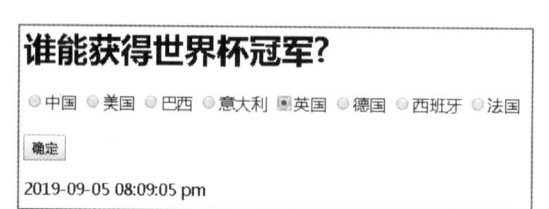

图 6-2

(3)结果页面:页头显示"投票结果页",页面内容显示用户名和所选择的国家名,页脚显示当前时间,如图 6-3 所示。

图 6-3

6.3 设计思路

1. 工程设计

创建 Dynamic Web Project 项目——Vote。

2. 文件设计

工程文件设计如表 6-1 所示。

表 6-1

类型	文件	说明
JSP 文件	WebContent/first.jsp	显示第一页
	WebContent/vote.jsp	显示投票页
	WebContent/result.jsp	显示结果页
	WebContent/footer.ssp	页脚页面
Java 文件	com.vote.servlet.StartServlet	开始投票处理类,响应/Start 请求
	com.vote.servlet.VoteServlet	投票处理类,响应/Vote 请求
css 文件	css/style.css	页面样式
配置文件	WebContent/WEB-INF/web.xml	Web 配置文件

3. 页面设计

(1)姓名输入页面如图 6-4 所示。

图 6-4

（2）投票页面如图 6-5 所示。

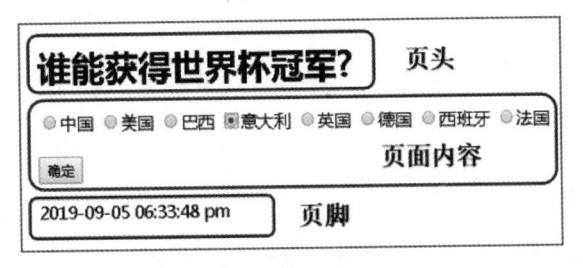

图 6-5

（3）结果页面如图 6-6 所示。

图 6-6

6.4　实验实施（跟我做）

步骤 1：创建工程

启动 Eclipse，选择"File"→"New"→"Dynamic Web Project"工程，在打开的向导窗口中的"Project name"文本框中输入"Vote"工程名，在"Target runtime"下拉列表中选择"Apache Tomcat v8.5"选项。单击"Next"按钮，在打开的"Web Module"配置窗口中选中"Generate web.xml deployment descriptor"复选框，向导将自动创建 web.xml 文件。

单击"Finish"按钮，完成创建工程的操作。在 Package Explorer 中可以看到新创建的 Vote 工程。

在"WebContent"文件夹下创建"css"文件夹，如图 6-7 所示。

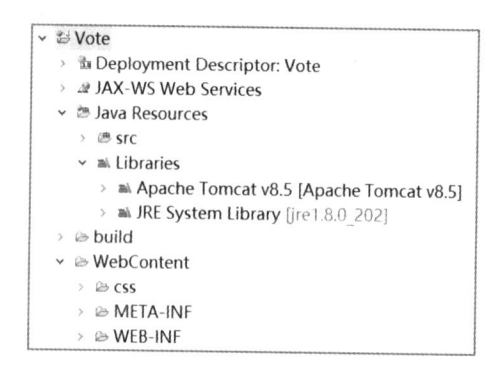

图 6-7

在"WebContent"文件夹下，分别创建 JSP 页面文件：first.jsp、footer.jsp、vote.jsp 和 result.jsp。

步骤 2：编写页面样式文件

打开"WebContent/css/style.css"文件，编写如下代码。

```css
body{
text-align: center;
}
h1{
text-align: center;
}
table{
width: 600px;
border: 1px solid #000000;
text-align: center;
margin: 0 auto;
}
th,td{
padding: 5px;
border: 1px solid #000000;
}
.button{
width: 280px;
margin: 0 2px;
}
span{
color: red;
}
```

步骤 3：编写公共 footer.jsp 页面

打开"WebContent"文件夹下的 footer.jsp 文件，清除自动生成的 HTML 代码，只保留第一行代码，使用 SimpleDateFormat 和 Date 类，在页面中输出当前日期和时间。

（1）修改页面的编码格式化为 UTF-8。

```
<%@ page language="java" contentType="text/html; charset=UTF-8"
    pageEncoding="UTF-8"%>
```

(2）导入 Date 和 SimpleDateFormat 类，导入 Locale 类。

```
<%@ page import="java.text.SimpleDateFormat"%>
<%@ page import="java.util.Date,java.util.Locale"%>
```

（3）嵌入 Java 代码，获得当前日期和时间的字符串。

```
<%
SimpleDateFormat sdf=new SimpleDateFormat("yyyy-MM-dd hh:mm:ss aa",
Locale.ENGLISH);
String curTime=sdf.format(new Date()).toLowerCase();
%>
```

Date 类默认创建的对象即为当前程序的时间，需要通过 SimpleDateFormat 类对象将 Date 对象按约定格式转换成字符串。其中，y 表示年，M 表示月，d 表示日，h 表示 1~12 制的小时数，m 表示分钟，s 表示秒，a 表示上/下午标识，默认显示中文的"上午"或"下午"，这里希望显示英文的"AM"或"PM"，则需要在 SimpleDateFormat 构造函数的第二个参数中设置为英文（Locale.ENGLISH），另外这里希望以小写显示，所以调用 String 类的 toLowerCase( )函数将字符串中的大写字母统一改为小写字母。

（4）输出 curTime 字符串。

```
<footer>
    <p><%=curTime%></p>
</footer>
```

步骤 4：配置用户名输入页面

1. 编写 first.jsp 页面，显示用户名输入表单

打开"WebContent"文件夹下的 first.jsp 文件，编写代码，页面分为页头、表单和页脚。页头使用<header>标签，页脚引用公共的 footer.jsp 页面。效果如图 6-4 所示。

（1）修改页面的编码格式化为 UTF-8。

```
<%@ page language="java" contentType="text/html; charset=UTF-8"
    pageEncoding="UTF-8"%>
```

（2）引入样式文件，设置标题为"输入用户名"。

```
<!DOCTYPE html>
<html>
<head>
<meta charset="UTF-8">
<link rel="stylesheet" type="text/css" href="css/style.css"/>
<title>输入用户名</title>
</head>
<body>
</body>
</html>
```

（3）编写页面标题。

```
<body>
    <header>
```

```
        <h1>输入用户名</h1>
    </header>
</body>
```

（4）编写表单。

```
<body>
    <header>
        <h1>输入用户名</h1>
    </header>
    <form action="" method="">
    <table>
      <tr>
        <td>用户名：</td>
        <td><input type="text" name="username"></td>
      </tr>
      <tr>
        <td colspan="2"><button type="submit">下一步</button></td>
      </tr>
    </table>
    </form>
</body>
```

（5）引入 footer.jsp 页脚页面。

```
<body>
    ……
    <%@ include file="footer.jsp" %>
</body>
```

2. 实现用户名提交功能

（1）设置表单提交方式为 POST，action 为"Start"。

```
<body>
    ……
    <form action="Start" method="POST">
    ……
    </form>
</body>
```

（2）在 src 包文件夹下创建 com.vote.servlet 包。

（3）在 com.vote.servlet 包中创建 Servlet 类 StartServlet，用来处理"/Start"请求。保留 doPost 方法，去掉 doGet 方法。

```
package com.vote.servlet;

import java.io.IOException;
import javax.servlet.ServletException;
import javax.servlet.annotation.WebServlet;
import javax.servlet.http.HttpServlet;
import javax.servlet.http.HttpServletRequest;
import javax.servlet.http.HttpServletResponse;
```

```java
/**
 * Servlet implementation class FirstServlet
 */
@WebServlet("/Start")
public class StartServlet extends HttpServlet{
    private static final long serialVersionUID=1L;

    /**
     * @see HttpServlet#HttpServlet()
     */
    public StartServlet(){
        super();
        // TODO Auto-generated constructor stub
    }



    /**
     * @see HttpServlet#doPost(HttpServletRequest request, HttpServletResponse response)
     */
    protected void doPost(HttpServletRequest request, HttpServletResponse response) throws ServletException, IOException{
    }

}
```

（4）编写 doPost 方法，直接将从表单中提交的 username 数据转发到下一个页面中，因为提交的内容可能包含中文，调用 setCharacterEncoding 方法，统一把编码格式设为"UTF-8"，将 username 值转发到 vote.jsp 页面。

```java
@WebServlet("/Start")
public class StartServlet extends HttpServlet{
    ......
    protected void doPost(HttpServletRequest request, HttpServletResponse response) throws ServletException, IOException{
        request.setCharacterEncoding("UTF-8");
        request.getRequestDispatcher("vote.jsp").forward(request, response);
    }

}
```

RequestDispatcher 代表请求的派发者。RequestDispatcher.forward(request, response)方法将请求从一个 Servlet or JSP 目标资源上转发到服务器上的另一个资源（servlet、JSP 文件或 HTML 文件，这些资源必须是当前 Web 上下文中的），让其他的资源去生成响应数据。

此处表单提交请求的是目标资源 StartServlet，StartServlet 接收到请求后，转发到 vote.jsp 页面，真正产生响应数据是被转发的资源 vote.jsp，而 StartServlet 只是起了引导转发作用。浏览器的地址栏不会变，依然是/Start 的 URL。这个方法可以允许被请求的目标资源做一些准备工作后,再让转发的资源去响应请求。这里可以将第一个页面输入的 username 值传到 vote.jsp 页面。

步骤 5:配置投票页面

1. 编写 vote.jsp 页面,显示投票表单

打开"WebContent"文件夹下的 vote.jsp 文件,编写代码,页面分为页头、表单和页脚。页头使用<header>标签,页脚引用公共的 footer.jsp 页面。效果如图 6-5 所示。

(1)修改页面的编码格式化为 UTF-8。

```
<%@ page language="java" contentType="text/html; charset=UTF-8"
    pageEncoding="UTF-8"%>
```

(2)引入样式文件,设置标题为"谁能获得世界杯冠军?"。

```
<!DOCTYPE html>
<html>
<head>
<meta charset="UTF-8">
<link rel="stylesheet" type="text/css" href="css/style.css"/>
<title>谁能获得世界杯冠军?</title>
</head>
<body>
</body>
</html>
```

(3)添加脚本获得转发来的 username 值。

```
<%
String username=request.getParameter("username");
%>
```

(4)编写页面标题。

```
<body>
    <header>
        <h1>谁能获得世界杯冠军?</h1>
    </header>
</body>
```

(5)编写表单,使用 hidden 隐藏域把 username 值保存下来。

```
<body>
    <header>
        <h1>谁能获得世界杯冠军?</h1>
    </header>
    <form action="" method="">
        <input type="hidden" name="username" value="<%=username%>">
        <div>
            <label><input type="radio" name="content" value="中国">中国</label>
            <label><input type="radio" name="content" value="美国">美国</label>
            <label><input type="radio" name="content" value="巴西">巴西</label>
            <label><input type="radio" name="content" value="意大利">意大利</label>
            <label><input type="radio" name="content" value="英国">英国</label>
```

```html
            <label><input type="radio" name="content" value="德国">德国</label>
            <label><input type="radio" name="content" value="西班牙">西班牙</label>
            <label><input type="radio" name="content" value="法国">法国</label>
        </div>
        <br/>
        <button type="submit">确定</button>
    </form>
</body>
```

(6) 引入 footer.jsp 页脚页面。

```html
<body>
    ……
    <%@ include file="footer.jsp"%>
</body>
```

2. 实现用户名提交功能

(1) 设置表单提交方式为 POST，action 为 "Vote"。

```html
<body>
    ……
    <form action="Vote" method="POST">
    ……
    </form>
</body>
```

(2) 在 com.vote.servlet 包中创建 Servlet 类 VoteServlet，用来处理 "/Vote" 请求。保留 doPost 方法，去掉 doGet 方法。

```java
package com.vote.servlet;

import java.io.IOException;
import javax.servlet.ServletException;
import javax.servlet.annotation.WebServlet;
import javax.servlet.http.HttpServlet;
import javax.servlet.http.HttpServletRequest;
import javax.servlet.http.HttpServletResponse;

/**
 * Servlet implementation class VoteServlet
 */
@WebServlet("/Vote")
public class VoteServlet extends HttpServlet{
    private static final long serialVersionUID=1L;

    /**
     * @see HttpServlet#HttpServlet()
     */
    public VoteServlet(){
        super();
        // TODO Auto-generated constructor stub
    }
```

```java
    /**
     * @see HttpServlet#doPost(HttpServletRequest request, HttpServletResponse response)
     */
    protected void doPost(HttpServletRequest request, HttpServletResponse response) throws ServletException, IOException{
    }
}
```

（3）编写 doPost 方法，直接将从表单中提交的 username 数据转发到下一个页面中，因为提交的内容可能包含中文，调用 setCharacterEncoding 方法，统一把编码格式设为"UTF-8"，将 username 和投票内容 content 转发到 result.jsp 页面。

```java
@WebServlet("/Vote")
public class VoteServlet extends HttpServlet{
    private static final long serialVersionUID=1L;

    /**
     * @see HttpServlet#HttpServlet()
     */
    public VoteServlet(){
        super();
        // TODO Auto-generated constructor stub
    }

    /**
     * @see HttpServlet#doPost(HttpServletRequest request, HttpServletResponse response)
     */
    protected void doPost(HttpServletRequest request, HttpServletResponse response) throws ServletException, IOException{
        request.setCharacterEncoding("UTF-8");
        request.getRequestDispatcher("result.jsp") .forward(request, response);
    }
}
```

步骤 6：数据的获取和写入

打开"WebContent"文件夹下的 result.jsp 文件，编写代码，页面分为页头、表单和页脚。页头使用<header>标签，页脚引用公共的 footer.jsp 页面。效果如图 6-6 所示。

（1）修改页面的编码格式化为 UTF-8。

```jsp
<%@ page language="java" contentType="text/html; charset=UTF-8"
    pageEncoding="UTF-8"%>
```

（2）引入样式文件，设置标题为"投票结果页"。

```html
<!DOCTYPE html>
```

```html
<html>
<head>
<meta charset="UTF-8">
<link rel="stylesheet" type="text/css" href="css/style.css"/>
<title>投票结果页</title>
</head>
<body>
</body>
</html>
```

(3) 添加脚本获得转发来的 username 和 content 值。

```jsp
<%
String username=request.getParameter("username");
String content=request.getParameter("content");
%>
```

(4) 编写页面标题。

```html
<body>
    <header>
        <h1>投票结果页</h1>
    </header>
</body>
```

(5) 编写表单,使用 hidden 隐藏域把 username 值保存下来。

```jsp
<body>
    <header>
        <h1>投票结果页</h1>
    </header>
    <%=username%> : <%=content%>
</body>
```

(6) 引入 footer.jsp 页脚页面。

```jsp
<body>
    ……
    <%@ include file="footer.jsp" %>
</body>
```

步骤 7:配置欢迎页面

打开"WebContent/WEB-INF"文件夹下的 web.xml 文件。修改代码,将 welcome-file 修改为 first.jsp。

```xml
<?xml version="1.0" encoding="UTF-8"?>
<web-app xmlns:xsi="http://www.w3.org/2001/XMLSchema-instance" xmlns="http://java.sun.com/xml/ns/javaee" xsi:schemaLocation="http://java.sun.com/xml/ns/javaee http://java.sun.com/xml/ns/javaee/web-app_3_0.xsd" id="WebApp_ID" version="3.0">
    <display-name>Vote</display-name>
    <welcome-file-list>
        <welcome-file>first.jsp</welcome-file>
    </welcome-file-list>
```

```
</web-app>
```

步骤 8：运行效果

（1）启动项目，或启动 Tomcat 后输入：http://localhost/Vote/first.jsp，显示"输入用户名"页面，输入用户名"user"，如图 6-1 所示。

（2）单击"下一步"按钮，进入"投票"页面，选中"英国"单选按钮，如图 6-2 所示。

（3）单击"确定"按钮，显示"投票结果面"页面，显示"user：英国"，页脚显示当前时间，如图 6-3 所示。

第 7 章
Java Web 编程（JSP+Servlet）：购物车

7.1 实验目标

（1）掌握 Java 的基础语法、编码规范。
（2）掌握 Java 包管理。
（3）掌握 Java 常量、变量的定义和使用方法。
（4）掌握 JSP 内置对象的使用方法。
（5）掌握 Java 基本运算符的使用方法。
（6）掌握 Java 条件判断语句的使用方法。
（7）掌握 Java 面向对象编程的方法。
（8）掌握 ArrayList 和 String 类的使用方法。
（9）掌握 Java 中启动 Session 的存取操作方法。
（10）综合应用 Java 的 Web 编程技术编写"购物车"程序。

7.2 实验任务

（1）在页面 index.jsp 中显示商品列表，每个商品后面购物车栏有一个 + 按钮，单击 + 按钮，可以向"我的购物车"中添加一个商品，如图 7-1 所示。在商品列表下面有一个"我的购物车"超链接，单击该超链接可以进入购物车页面 cart.jsp。

（2）通过 index.jsp 商品列表页面的"我的购物车"超链接，进入购物车页面 cart.jsp，如图 7-2 所示。

在购物车页面中可以更改购买数量，单击"结算"按钮，计算购物车内的商品总价，进入确认订单页面 order.jsp。

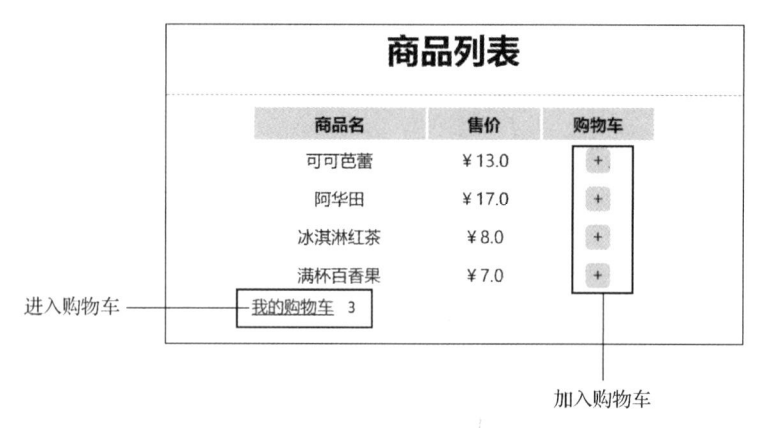

图 7-1

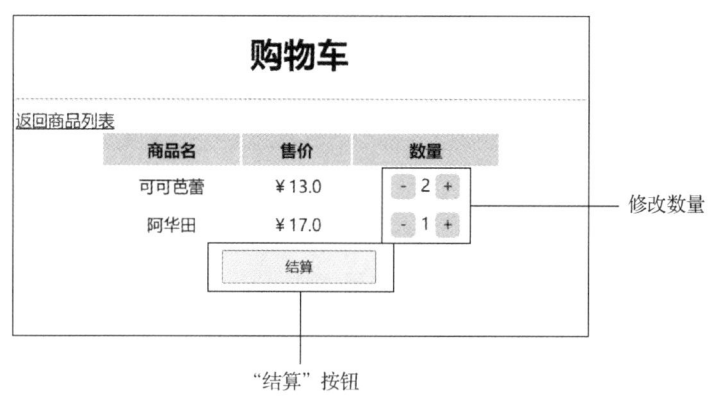

图 7-2

（3）在确认订单页面 order.jsp，显示各商品的单价和数量，以及商品总量和总价，在地址输入框中填写订单地址，如图 7-3 所示，单击"提交订单"按钮，生成订单进入订单页面 done.jsp。

图 7-3

（4）订单页面 done.jsp 显示地址信息和所购买的商品信息，如图 7-4 所示。

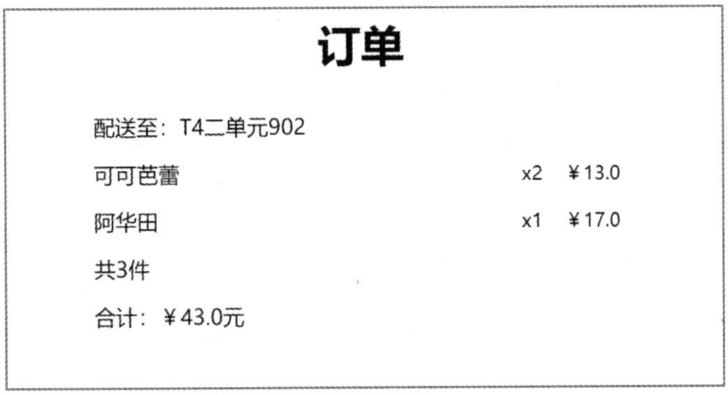

图 7-4

7.3 设计思路

1．工程设计

创建 Dynamic Web Project 项目——ShopingCart。

2．程序结构

程序目录结构如图 7-5 所示。

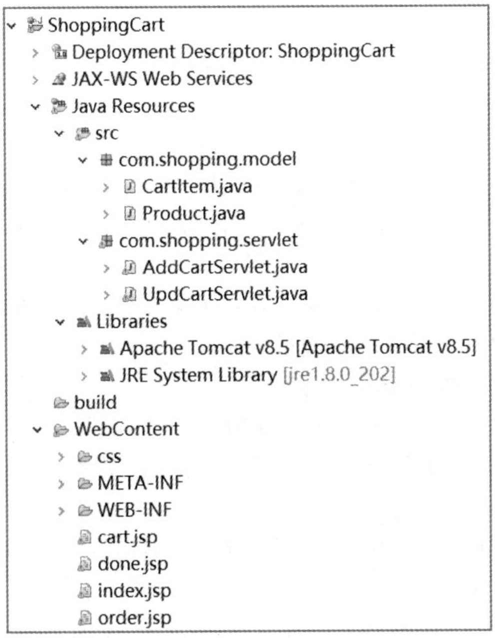

图 7-5

3．文件设计

文件设计如表 7-1 所示。

表 7-1

类型	文件/类	说明
JSP 文件	index.jsp	商品列表页面
	cart.jsp	购物车页面
	order.jsp	确认订单页面
	done.jsp	订单页面
Java 类	com.shopping.servlet.AddCartServlet	添加购物车请求处理类
	com.shopping.servlet.UpdCartServlet	修改购物车请求处理类
	com.shopping.model.Product	商品实体类
	com.shopping.model.CartItem	购物车商品项类
css 文件	css/style.css	全部样式

4．请求 URL 设计

请求 URL 设计如表 7-2 所示。

表 7-2

URL 格式	Servlet 类	说明
/AddCart	AddCartServlet	向购物车中添加一个商品
/UpdCart	UpdCartServlet	调整购物车中商品的数量

5．商品列表页面 index.jsp

（1）单个商品包含 id、name、price 3 个字段，创建 Product 商品类保存 1 个商品的数据，使用 ArrayList 数组保存多个商品列表。

（2）使用 for( )循环遍历列表，将商品信息以表格形式显示。

（3）单击 + 按钮，进行加入购物车操作；请求 URL：/AddCart，将商品的 id、name、price 作为参数传递到 AddCartServlet::doGet 方法处理，判断 Session 变量中的"cart"对象是否为空，若为空则执行以下操作。

① 创建 List<CartItem>类对象 order，用来保存添加到购物车中的商品列表。

② 创建 CartItem 对象，设置商品数量为 1。

③ 将 CartItem 对象添加到 order 中。

④ 将 order 对象添加到 Session 中，key 值设为"cart"。

若不为空，则判断 Session 变量中是否已有该商品，没有就新增一个商品，已有就使该商品数量加 1。

（4）单击"我的购物车"超链接，将页面跳转到购物车页面 cart.jsp。

6．购物车页面 cart.jsp

（1）获取 Session 中的购物车信息，单击 - 按钮或 + 按钮更改购物车内的商品数量，更新 Session 中的"cart"对象。

（2）使用 for 循环遍历 order 列表，对"num"属性值求和，得到购物车内的商品总量，存入变量 Session 中的"num"属性中。

（3）计算购物车内的商品总价，存入 Session 中的"sum"属性中。

（4）单击"结算"按钮，进入确认订单页面 order.jsp。

7. 确认订单页面 order.jsp

（1）获取 Session 中的"cart"、"num"和"sum"对象，以列表形式显示购物车内的商品单价、数量、总价和总量。

（2）在地址输入框中填写地址后，单击"提交订单"按钮，进入订单页面 done.jsp。

8. 订单页面 done.jsp

（1）使用 request.getParamter( ) 获取表单提交的地址信息，显示在页面中。

（2）显示购物车内的商品单价、数量、总价和总量。

7.4 实验实施（跟我做）

步骤 1：制作商品列表页面

1. 创建 css/style.css 文件

添加 css/style.css 文件的代码如下：

```css
h1,p{
  text-align: center;
}

table{
  width: 70%;
  text-align: center;
  margin: 0 auto;
}

th{
  background: #dddddd;
}

th,td{
  padding: 5px;
}

div{
  width: 70%;
  margin: 0 auto;
}

a.btn{
  display: inline-block;
  width: 1.5em;
  height: 1.5em;
  background: #dddddd;
  text-decoration: none;
```

```css
  color: black;
  border-radius: 5px;
}

input{
  width: 150px;
  padding: 5px 0;
}

span{
  font-size: 14px;
  padding: 0 8px;
}

ul{
  width: 400px;
  margin: 0 auto;
  padding: 0;
  list-style: none;
}

li{
  line-height: 35px;
}

li span{
  float: right;
}

.address, .address input{
  display: block;
  text-align: center;
  margin: 0 auto;
}
```

2. 创建实体类

（1）创建 com.shopping.model 包。

（2）新建 Product 和 CartItem 类。

（3）编写 Product 类，代码如下：

```java
package com.shopping.model;

/**
 * 产品
 */
public class Product{
  private int id;        //id标识符
  private String name;   //商品名称
```

```java
    private double price;  //单价

    public Product(int id, String name, double price){
      super();
      this.id=id;
      this.name=name;
      this.price=price;
    }

    public int getId(){
      return id;
    }

    public void setId(int id){
      this.id=id;
    }

    public String getName(){
      return name;
    }

    public void setName(String name){
      this.name=name;
    }

    public double getPrice(){
      return price;
    }

    public void setPrice(double price){
      this.price=price;
    }
}
```

（4）编写 CartItem 类，代码如下：

```java
package com.shopping.model;
/**
 * 购物车商品项
 */
public class CartItem{
  private Product product;   //商品
  private int num;           //数量

  public CartItem(Product product, int num){
    super();
    this.product=product;
    this.num=num;
```

```java
    }

    public Product getProduct(){
        return product;
    }

    public void setProduct(Product product){
        this.product=product;
    }

    public int getNum(){
        return num;
    }

    public void setNum(int num){
        this.num=num;
    }
}
```

3. 创建商品列表页面 index.jsp

（1）在页面头部定义商品信息列表。

```jsp
<%@ page language="java" contentType="text/html; charset=UTF-8"
    pageEncoding="UTF-8"%>
<%@page import="java.util.List,java.util.ArrayList"%>
<%@page import="com.shopping.model.*"%>
<%
// 初始化商品列表
List<Product> products=new ArrayList<Product>();
Product p1=new Product(1, "可可芭蕾", 13.0);
products.add(p1);
Product p2=new Product(2, "阿华田", 17.0);
products.add(p2);
Product p3=new Product(3, "冰淇淋红茶", 8.0);
products.add(p3);
Product p4=new Product(4, "满杯百香果", 7.0);
products.add(p4);
```

（2）从 Session 中获得购物车商品列表，统计购物车内的商品数量。

```jsp
//计算购物车中的商品数量
int sum=0;
if(session.getAttribute("cart")!=null){
// 获得 session 中购物车商品列表
    List<CartItem>order=(List<CartItem>)session.getAttribute("cart");
    for(CartItem item : order){
        sum+=item.getNum(); //累加商品数量
    }
}
```

```
%>
```

（3）显示商品信息，以及购物车内的商品总量。

```html
<!DOCTYPE html>
<html>
<head>
<meta charset="UTF-8">
<link rel="stylesheet" type="text/css" href="css/style.css"/>
<title>购物网站</title>
</head>
<body>
  <div>
    <h1>商品列表</h1>
    <hr>
    <table>
      <tr>
        <th>商品名</th>
        <th>售价</th>
        <th>购物车</th>
      </tr>
      <!-- 遍历商品数组,显示商品信息 -->
      <%
      for (Product product : products){
      %>
      <tr>
        <td><%=product.getName()%></td>
        <!-- 商品名 -->
        <td>¥<%=product.getPrice()%></td>
        <!-- 商品价格 -->
        <td><a class="btn"
           href="AddCart?upd=add&id=<%=product.getId()%>&name=<%=product.getName()%>&price=<%=product.getPrice()%>">+</a></td>
      </tr>
      <%
      }
      %>
    </table>
    <!-- 显示购物车内商品总数 -->
    <div>
      <a href="AddCart?upd=cart">我的购物车</a> <span> <%=sum%>
      </span>
    </div>
  </div>
</body>
</html>
```

步骤 2：将商品加入购物车

（1）在 com.shopping.servlet 包中新建 HttpServlet 子类 AddCartServlet 类，添加 doGet 函数。使用@WebServlet 注解设置响应该 Servlet 的路径"/AddCart"。

```java
package com.shopping.servlet;

import java.io.IOException;
import javax.servlet.ServletException;
import javax.servlet.annotation.WebServlet;
import javax.servlet.http.HttpServlet;
import javax.servlet.http.HttpServletRequest;
import javax.servlet.http.HttpServletResponse;

/**
 * Servlet implementation class AddCartServlet
 */
@WebServlet("/AddCart")
public class AddCartServlet extends HttpServlet{
  private static final long serialVersionUID=1L;

  /**
   * @see HttpServlet#HttpServlet()
   */
  public AddCartServlet(){
    super();
    // TODO Auto-generated constructor stub
  }

  /**
   * @see HttpServlet#doGet(HttpServletRequest request, HttpServletResponse
   * response)
   */
  protected void doGet(HttpServletRequest request, HttpServletResponse response)
      throws ServletException, IOException{
    //
  }
}
```

（2）在 web.xml 文件中添加该 servlet。

```xml
<?xml version="1.0" encoding="UTF-8"?>
<web-app xmlns:xsi="http://www.w3.org/2001/XMLSchema-instance"
  xmlns="http://java.sun.com/xml/ns/javaee"
  xsi:schemaLocation="http://java.sun.com/xml/ns/javaee http://java.sun.com/xml/ns/javaee/web-app_3_0.xsd" id="WebApp_ID" version="3.0">
```

```xml
  <display-name>ShoppingCart</display-name>

  <servlet>
    <servlet-name>AddCart</servlet-name>
    <servlet-class>com.shopping.servlet.AddCartServlet</servlet-class>
  </servlet>

  <welcome-file-list>
    <welcome-file>index.jsp</welcome-file>
  </welcome-file-list>
</web-app>
```

（3）获得操作码"upd"，判断操作码为"add"时，将商品加入购物车。

判断 Session 中是否包含 cart 变量，若不包含，则将商品信息存入数组，创建 Session 变量。若 Session 存在，则判断购物车中是否已有该商品。

```java
//获得upd操作码
String upd=request.getParameter("upd");
//获得Session
HttpSession session=request.getSession();
//添加商品到购物车中
if(upd.equals("add")){
  //获得商品信息
  int id=Integer.parseInt(request.getParameter("id"));
  String name=request.getParameter("name");
  double price=Double.parseDouble(request.getParameter("price"));
  //判断购物车中是否已添加商品
  if(session.getAttribute("cart")==null){//未添加
    List<CartItem>order=new ArrayList<CartItem>();
    CartItem item=new CartItem(new Product(id, name, price), 1);
    order.add(item);
    session.setAttribute("cart", order);
  }else{//已添加
    List<CartItem>order=(List<CartItem>) session.getAttribute("cart");
    boolean hasProduct=false; //标识要添加的商品是否已在购物车中
    for(CartItem item : order){
      if(item.getProduct().getId()==id){
        item.setNum(item.getNum()+1);
        hasProduct=true;
        break;
      }
    }
    if(hasProduct==false){
      CartItem item=new CartItem(new Product(id, name, price), 1);
      order.add(item);
      session.setAttribute("cart", order);
    }
  }
}
```

```
    //跳转到主页
    response.sendRedirect("index.jsp");
}
```

（4）操作码为"cart"时，如果 Session 中的 cart 变量不为空，则离开商品页面，进入购物车页面。

```
if(upd.equals("cart")){
  if(session.getAttribute("cart")!=null){
    response.sendRedirect("cart.jsp");
  } else{
    response.sendRedirect("index.jsp");
  }
}
```

页面效果如图 7-6 所示。

图 7-6

步骤 3：制作购物车页面

（1）创建购物车页面 cart.jsp。
（2）从 Session 中获取购物车的商品信息和操作码。
（3）显示购物车内的商品信息。

```
<%@ page language="java" contentType="text/html; charset=UTF-8"
  pageEncoding="UTF-8"%>
<%@page import="java.util.List,java.util.ArrayList"%>
<%@page import="com.shopping.model.*"%>
<%
List<CartItem> order=(List<CartItem>) session.getAttribute("cart");
%>
<!DOCTYPE html>
<html>
<head>
<meta charset="UTF-8">
<link rel="stylesheet" type="text/css" href="css/style.css"/>
<title>购物网站</title>
</head>
```

```html
<body>
  <div>
    <h1>购物车</h1>
    <hr>
    <a href="index.jsp">返回商品列表</a>
    <table>
      <tr>
        <th>商品名</th>
        <th>售价</th>
        <th>数量</th>
      </tr>
      <%
      for(CartItem item : order){
      %>
      <tr>
        <!-- 商品名-->
        <td><%=item.getProduct().getName()%></td>
        <!-- 商品价格-->
        <td>¥<%=item.getProduct().getPrice()%></td>
            <!-- 商品数量-->
        <td>
           <a class="btn" href="UpdCart?upd=0&id=<%=item.getProduct().getId()%>">-</a>
                <%=item.getNum()%>
           <a class="btn" href="UpdCart?upd=1&id=<%=item.getProduct().getId()%>">+</a>
        </td>
      </tr>
      <% } %>
      <tr>
        <td colspan="3">
          <form action="UpdCart" method="get">
            <input type="submit" value="结算">
          </form>
        </td>
      </tr>
    </table>
  </div>
</body>
</html>
```

步骤4：改变购物车内的商品数量

（1）创建 com.shopping.servlet.UpdCartServlet 类，编写 doGet( )函数，实现购物车修改请求。

（2）更改商品数量，计算商品总价，跳转到确认订单页面。

```java
protected void doGet(HttpServletRequest request, HttpServletResponse response)
    throws ServletException, IOException{
  //操作码
  String upd=request.getParameter("upd");
  //获得商品id
  int id=0;
  if(request.getParameter("id")!=null){
    id=Integer.parseInt(request.getParameter("id"));
  }
  //获得Session中的购物车商品列表
  HttpSession session=request.getSession();
  List<CartItem> order=(List<CartItem>) session.getAttribute("cart");
  Double sum=new Double(0);        //购物车总价格
  Integer num=new Integer(0);      //购物车中的商品总数
  CartItem delItem=null;           //标识要删除的商品项
  //遍历订单中的商品项
  for(CartItem item: order){
    if(upd==null){
      //计算商品总价格
      sum+=item.getNum()*item.getProduct().getPrice();
    } else{
      if(item.getProduct().getId()==id){
        switch (upd){
        case "0": //减少商品数量
          if(item.getNum()>1){
            item.setNum(item.getNum()-1);
          }else{
            item.setNum(0);
            delItem=item;
          }
          break;
        case "1": //增加商品数量
          item.setNum(item.getNum()+1);
          break;
        default:
          break;
        }
      }
    }
    num+=item.getNum();
  }
  //判断是否有删除的商品项
  if(delItem!=null){
    order.remove(delItem);
  }
  //将订单、总价格和总数量添加到session中
```

```
    session.setAttribute("cart", order);
    session.setAttribute("sum", sum);
    session.setAttribute("num", num);

    if(upd==null){
      response.sendRedirect("order.jsp");
    }else{
      response.sendRedirect("cart.jsp");
    }
}
```

页面效果如图 7-7 所示。

图 7-7

步骤 5：制作确认订单页面

（1）创建确认订单页面 order.jsp。
（2）从 Session 获取购物车内的商品信息、商品总量和商品总价。
（3）显示商品信息。

```
<%@ page language="java" contentType="text/html; charset=UTF-8"
    pageEncoding="UTF-8"%>
<%@page import="java.util.List,java.util.ArrayList"%>
<%@page import="com.shopping.model.*"%>
<%
Integer num=(Integer) session.getAttribute("num");
Double sum=(Double) session.getAttribute("sum");
List<CartItem>order=(List<CartItem>)session.getAttribute("cart");
%>
<!DOCTYPE html>
<html>
<head>
<meta charset="UTF-8">
<link rel="stylesheet" type="text/css" href="css/style.css"/>
<title>购物网站</title>
</head>
<body>
```

```html
<div>
    <h1>确认订单</h1>
    <hr>
    <a href="index.jsp">返回商品列表</a>
    </p>
    <table>
      <!-- 商品信息 -->
      <%
      for(CartItem item : order){
      %>
      <tr>
        <td><%=item.getProduct().getName()%></td>
        <!-- 商品名 -->
        <td>¥<%=item.getProduct().getPrice()%><!-- 商品单价 --> <span>x<%=item.getNum()%></span>
          <!-- 商品数量 -->
        </td>
      </tr>
      <%
      }
      %>

    </table>
  </div>
</body>
</html>
```

（4）显示商品总量和商品总价。

```html
<tr>
  <td colspan="2"><span>共<%=num%>件
    </span> <!-- 商品总量 --> 小计：¥<%=sum%> <!-- 商品总价 --></td>
</tr>
```

（5）订单地址输入框。

```html
<!-- 订单地址 -->
<form action="done.jsp" method="post" class="address">
  <textarea name="address" placeholder="输入地址" cols="60" required></textarea>
  <input type="submit" value="提交订单">
</form>
```

页面效果如图 7-8 所示。

图 7-8

步骤 6：制作订单页面

（1）创建订单页面 done.jsp。
（2）从 Session 获取购物车内的商品信息、商品总量和商品总价。
（3）获取并显示在确认订单页面中输入的订单地址。

```jsp
<%@ page language="java" contentType="text/html; charset=UTF-8"
    pageEncoding="UTF-8"%>
<%@page import="java.util.List,java.util.ArrayList"%>
<%@page import="com.shopping.model.*"%>
<%
//获取 session 中的订单、总价格和总数量
Integer num=(Integer)session.getAttribute("num");
Double sum=(Double)session.getAttribute("sum");
List<CartItem>order=(List<CartItem>) session.getAttribute("cart");
//获得 request 中的地址信息
String address=request.getParameter("address");
%>
<!DOCTYPE html>
<html>
<head>
<meta charset="UTF-8">
<link rel="stylesheet" type="text/css" href="css/style.css"/>
<title>购物网站</title>
</head>
<body>
  <h1>订单</h1>
  <ul>
    <li>配送至：<%=address %></li>
    <!-- 订单地址 -->
    <% for(CartItem item : order){ %>
    <li><%=item.getProduct().getName() %><!-- 商品名 --> <span>¥<%=item.getProduct().getPrice() %></span>
        <!-- 商品单价 --><span>x<%=item.getNum()%></span> <!-- 商品数量 -->
    </li>
```

```
        <% } %>
        <li>共<%=num %>件</li>
        <!-- 商品总量 -->
        <li>合计：¥<%=sum %></li>
        <!-- 商品总价 -->
    </ul>
</body>
</html>
```

页面效果如图 7-9 所示。

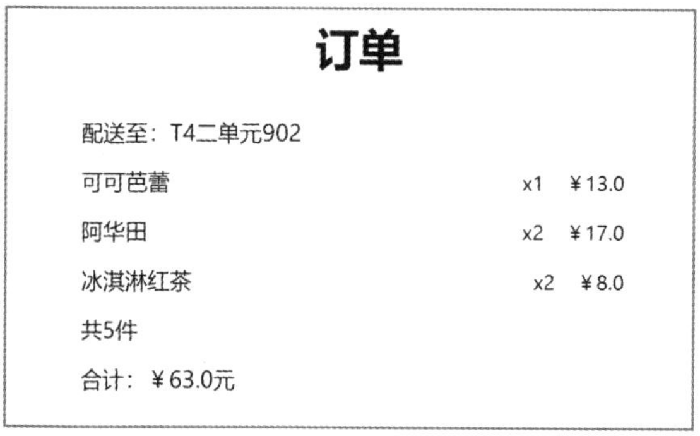

图 7-9

第 8 章 MySQL 数据库：MySQL 基本操作

8.1 实验目标

（1）掌握 MySQL 下载、安装及登录的方法。
（2）掌握 MySQL 创建数据库命令。
（3）掌握 MySQL 进入数据库命令。
（4）掌握 MySQL 查看数据库命令。
（5）掌握 MySQL 删除数据库命令。
（6）综合应用 MySQL 数据库软件，进行"MySQL 数据库基本操作"。

8.2 实验任务

（1）下载并按照 XAMPP 集成开发环境，使用 XAMPP 集成开发环境中的 MySQL 数据库。
（2）使用 XAMPP 控制面板中的"Shell"命令行工具修改 MySQL 登录密码并登录 MySQL 数据库。
（3）在浏览器中打开 XAMPP 中的 MySQL Web 管理端界面，使用 MySQL Web 管理端登录数据库。
（4）使用数据库的常用命令。
① 使用"create database"命令创建 questiondb 数据库。
② 使用"use"命令进入 questiondb 数据库。
③ 使用"show databases"命令查看 questiondb 数据库。

8.3 设计思路

1. 下载和安装 MySQL

下载并按照 XAMPP 集成开发环境，使用 XAMPP 集成开发环境中的 MySQL 数据库。

2．使用命令行登录 MySQL

（1）打开 XAMPP 控制面板中的"Shell"命令行工具。
（2）使用 MySQL 命令修改登录密码。
（3）使用 MySQL 命令登录数据库。

3．使用 MySQL Web 管理端登录数据库

（1）单击控制面板 Apache 中的 Config，打开 phpMyAdmin(config.inc.php)文件，修改 config.inc.php 中的 MySQL 密码，以便能够操作 MySQL Web 管理端。
（2）在浏览器中打开 MySQL 管理界面，登录 MySQL。

4．使用数据库中的常用命令

（1）在 MySQL 命令行中使用"create database"命令创建 questiondb 数据库。
（2）在 MySQL 命令行中使用"use"命令进入 questiondb 数据库。
（3）在 MySQL 命令行中使用"show databases"命令查看 questiondb 数据库。

8.4 实验实施（跟我做）

步骤 1：下载和安装 MySQL 数据库

1．下载 MySQL 数据库

有以下两种方法可以下载 MySQL 数据库。
方法 1：进入 MySQL 官网（https://dev.mysql.com/downloads/mysql/）进行下载。
方法 2：下载 XAMPP 集成套件，XAMPP 集成套件中包括 MySQL 数据库。

2．下载 XAMPP

进入 XAMPP 官方网站（https://www.apachefriends.org/zh_cn/download.html）进行下载，如图 8-1 所示。

图 8-1

3．打开 XAMPP Control Panel

打开 XAMPP Control Panel 软件，我们可以看到有一项是 MySQL，如图 8-2 所示。

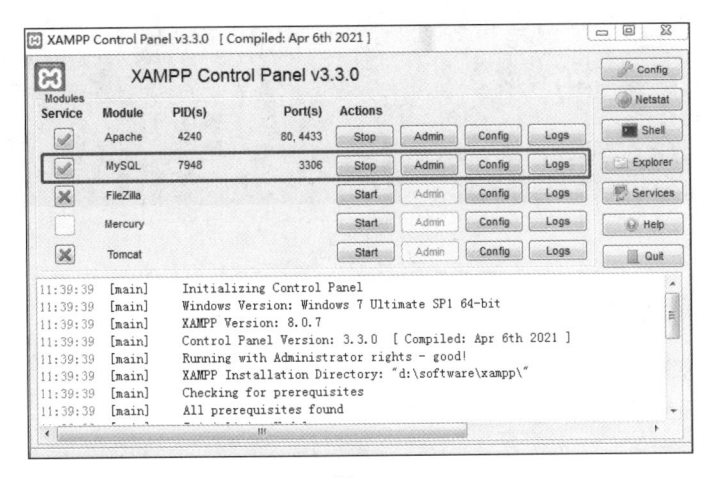

图 8-2

步骤 2：登录 MySQL 数据库

XAMPP 中有以下两种方法可以登录 MySQL 数据库。

（1）单击 XAMPP 控制面板中的"Shell"按钮，如图 8-3 所示，打开 Windows 命令提示符。

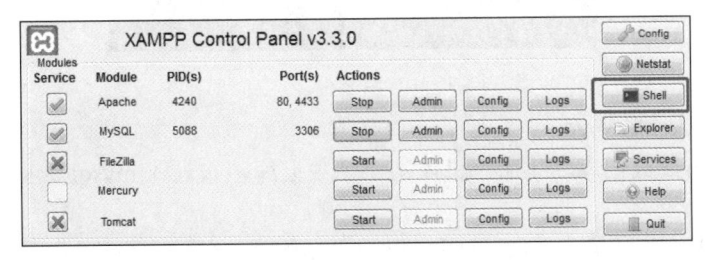

图 8-3

（2）使用以下语法修改 MySQL 密码，如新密码设置为 123456，如图 8-4 所示。

```
# mysqladmin --user=root password 123456
```

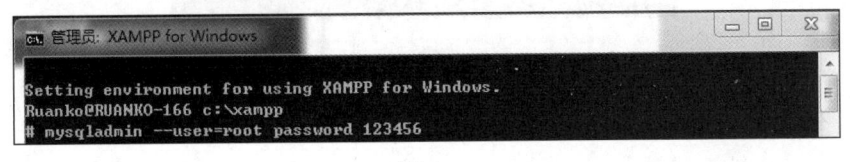

图 8-4

（3）执行命令"mysql -u root –p"，这时会提示输入密码，输入正确密码即可登录成功，进入 MySQL 命令模式，如图 8-5 所示。

图 8-5

步骤 3：数据库基本指令

（1）查看所有数据库

可以使用"show databases"命令查看当前 MySQL 包含哪些数据库。输入如下命令：
```
show databases;
```
效果如图 8-6 所示。

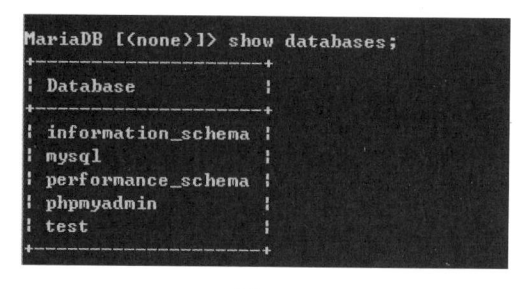

图 8-6

（2）进入 mysql 数据库

可以使用"use 数据库名"命令进入某一个数据库，如进入 mysql 数据库，则输入如下的命令：
```
use mysql;
```
效果如图 8-7 所示。

图 8-7

步骤 4：编写 SQL

在命令行中编写 SQL 语句来测试 MySQL 数据库。

（1）创建一个数据库 questiondb

使用 create database 语句创建数据库，语句如下：
```
create database questiondb;
```
效果如图 8-8 所示。

```
MariaDB [mysql]> create database questiondb;
Query OK, 1 row affected (0.005 sec)
```

图 8-8

（2）查看数据库

使用 show 命令查看 questiondb 数据库是否创建成功，语句如下：

```
show databases;
```

效果如图 8-9 所示。

```
MariaDB [mysql]> show databases;
+--------------------+
| Database           |
+--------------------+
| information_schema |
| mysql              |
| performance_schema |
| phpmyadmin         |
| questiondb         |
| test               |
+--------------------+
```

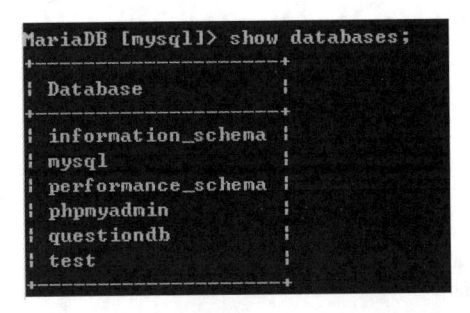

图 8-9

（3）删除数据库

使用 drop database 语句删除数据库，语句如下：

```
drop database [if exists] questiondb;
```

第 9 章
MySQL 数据库：试题信息管理

9.1 实验目标

(1) 掌握 MySQL 数据库中创建数据库的方法。
(2) 掌握 MySQL 数据库中创建表、设置约束、设置自增型字段的方法。
(3) 掌握 MySQL 数据库中修改表、删除表的方法。
(4) 掌握 MySQL 数据库中表数据的插入、修改、查询和删除方法。
(5) 掌握 MySQL 数据库存储过程的使用方法。
(6) 掌握 MySQL 数据库触发器的使用方法。
(7) 掌握 MySQL 数据库事务的使用方法。
(8) 掌握 MySQL 数据库视图的使用方法。
(9) 综合应用 MySQL 数据库软件，进行"试题信息管理"数据库各类操作和管理。

9.2 实验任务

试题管理系统是用于试题输入的管理系统，试题管理系统数据库则用于对系统中的试题等相关数据进行存储和管理，包括试题表 t_question 和选项表 t_option，一个试题包含多个选项，因此试题与选项的关系是一对多的关系，数据库关系如图 9-1 所示。

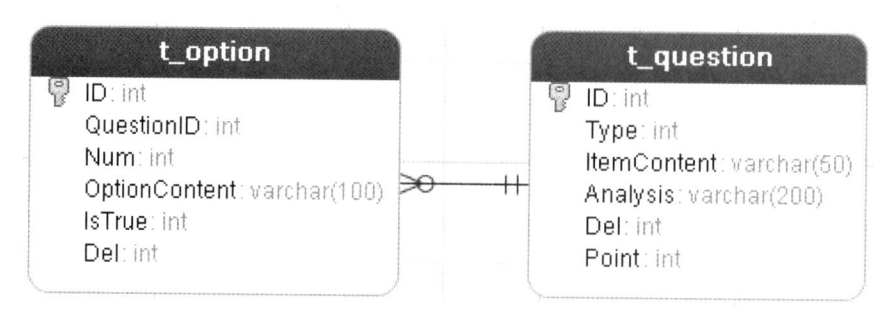

t_option
- ID: int
- QuestionID: int
- Num: int
- OptionContent: varchar(100)
- IsTrue: int
- Del: int

t_question
- ID: int
- Type: int
- ItemContent: varchar(50)
- Analysis: varchar(200)
- Del: int
- Point: int

图 9-1

试题表 t_question 的详细信息如表 9-1 所示。

表 9-1

名称	字段名	数据类型	备注
ID	ID	int(11)	主键，自增，每次增量为 1
题型	Type	int(11)	0 表示单选，1 表示多选，不能为空
题干	ItemContent	varchar(50)	不能为空
题典分析	Analysis	varchar(200)	可以为空
删除标志	Del	int(11)	0 表示正常，1 表示已删除，默认值为 0
分值	Point	int(11)	试题的分值，默认值为 0

选项表需要与试题表进行关联，选项表 t_option 的详细信息如表 9-2 所示。

表 9-2

名称	字段名	数据类型	备注
ID	ID	int(11)	主键，自增，每次增量为 1，不能为空
试题 ID	QuestionID	int(11)	外键，参照试题表，不能为空
选项号	Num	int(11)	0 表示 A，1 表示 B，2 表示 C，3 表示 D，不能为空
选项内容	OptionContent	varchar(100)	不能为空
标示是否是正确选项	IsTrue	int(11)	0 表示不是正确选项，1 表示是正确选项，不能为空
删除标识	Del	int(11)	0 表示正常，1 表示已删除，默认值为 0

本章需完成如下试题管理系统的数据库操作。

（1）创建试题管理系统数据库 questiondb。
（2）创建试题表 t_question 和选项表 t_option。
（3）对试题表 t_question 进行表的修改操作。
① 向试题表增加"分数"属性，数据类型为 float，默认值为'0'。
② 修改字段类型，将 Score 的数据类型改为整数。
③ 修改字段名，将 Score 改为 Point。
④ 复制试题表 t_question 的结构，创建 t_questionCopy 表。
⑤ 对试题表 t_questionCopy1 进行表的删除操作。
（4）为试题表 t_question 创建一个虚拟表 v_question，作为试题表的一个视图，它只显示试题表中的 ID 和题干信息。
（5）为 t_question 表的 ID 列按降序创建一个唯一索引。
（6）向试题表 t_question 中插入一条记录。
（7）将创建表结构和数据的 SQL 语句生成 SQL 脚本。
（8）开启事务，并将 t_question 表中 ID 为 1 的记录的 Del 字段值修改为 1，之后进行事务回滚。
（9）创建触发器，使选项表与相应试题表的删除标志 Del 同步更新。
（10）创建一个存储过程，新增一道试题，向试题表和选项表分别添加一条记录。

9.3 设计思路

完成试题管理系统数据库创建、数据表创建、数据表修改、视图创建、索引创建、事务管理、触发器创建、存储过程创建等操作。

1. 创建数据库和表

（1）创建试题管理系统数据库 questiondb。

```
create database questiondb;
```

（2）创建试题表 t_question。

```
create table t_question
(
ID int not null primary key auto_increment,
Type int not null check(Type IN(0,1)),
ItemContent varchar(50) not null,
Analysis varchar(200),
Del int not null default 0 check(Del IN(0,1))
);
```

（3）创建选项表 t_option。

```
create table t_option
(
ID int not null primary key auto_increment,
QuestionID int not null,
Num int not null check(Num IN(0,1,2,3)),
OptionContent varchar(100) not null,
IsTrue int not null check(IsTrue IN(0,1)),
Del int not null default 0 check(Del IN(0,1)),
foreign key(QuestionID) references t_question(ID)
);
```

2. 创建试题表的视图

创建 t_question 表的视图，只显示 ID 列与 ItemContent 列。

```
create view v_question AS select ID,ItemContent from t_question;
```

3. 为 t_question 表的 ID 列按降序创建一个唯一索引

```
create unique INDEX QueID on t_question(ID) DESC;
```

4. 管理表数据

对试题表 t_question 进行表的修改操作。

（1）向试题表增加"分数"属性，数据类型为 float，默认值为'0'。

```
alter table t_question add Score float(3,1) default '0';
```

（2）修改字段类型，将 Score 的数据类型改为整数。

```
alter table t_question modify column Score int;
```

（3）修改字段名，将 Score 改为 Point。

```
alter table t_question change Score Point int default '0';
```

（4）复制试题表 t_question 结构，创建 t_questionCopy 表。

```
CREATE TABLE t_questionCopy (LIKE t_question);
```

（5）删除 t_questionCopy 表。
```
drop table t_questionCopy cascade;
```
（6）向试题表 t_question 中插入一条记录。
```
insert into t_question(ID,Type,ItemContent,Analysis,Del)
values(1,0,'数据库系统的核心是','数据库知识',0);
```
（7）向选项表 t_option 中插入 4 条记录。
```
insert into t_option values('1','1','0','数据模型','1','0'), ('2','1','1','数据库管理系统','0','0'), ('3','1','2','数据库','0','0'), ('4','1','3','数据库管理员','0','0');
```

5．查询表数据

（1）查询 t_question 表中的 ID 列与 ItemContent 列。
```
select ID,ItemContent from t_question;
```
（2）查询 t_question 表中的所有列。
```
select * from t_question;
select ID,Type,ItemContent,Analysis,Del,Point from t_question;
```

6．备份以还原的方法

（1）将数据库 questiondb 导出到"D:/mysql_study"文件夹中。
```
mysqldump -uroot -p --default-character-set=utf8 questiondb > D:/mysql_study/questiondb_export.sql
```
（2）进入 questiondb 数据库并导入"D:/mysql_study/questiondb.sql"位置的脚本。
```
source D:/mysql_study/questiondb.sql;
```

7．事务控制

开启事务，并将 t_question 表中 ID 为 1 的记录的 Del 字段值修改为 1，之后进行事务回滚，查看修改结果。
```
start transaction;
update t_question set Del=1 where ID=1;
rollback;
select*from t_question;
```

8．添加触发器

（1）为了使选项表与相应试题表的删除标志 Del 同步更新，创建如下触发器。
```
create trigger shanchu after update on t_question for each row
update t_option set Del=(select Del from t_question where t_option.QuestionID=t_question.ID);
```
（2）通过 UPDATE 语句激活触发器，更新试题表，设置 ID 为 1 的记录的 Del 值为 1。
```
update t_question set Del=1 where ID=1;
```

9．创建存储过程

创建一个存储过程，向试题库新增一道试题，向试题表和选项表分别添加一条记录。

9.4 实验实施（跟我做）

步骤 1：创建数据库

（1）创建一个试题管理系统数据库 questiondb，如图 9-2 所示。

```
create database questiondb;
```

```
mysql> create database questiondb;
Query OK, 1 row affected (0.00 sec)
```

图 9-2

（2）进入刚才创建的 questiondb 数据库，如图 9-3 所示。

```
use questiondb;
```

```
mysql> use questiondb;
Database changed
```

图 9-3

（3）查看所有数据库，如图 9-4 所示。

```
SHOW DATABASES;
```

```
mysql> show databases;
+--------------------+
| Database           |
+--------------------+
| information_schema |
| mysql              |
| performance_schema |
| questiondb         |
| test               |
+--------------------+
5 rows in set (0.00 sec)
```

图 9-4

步骤 2：创建表

（1）在当前数据库中创建一个试题表 t_question。

```
create table t_question
(
ID int not null primary key auto_increment,
Type int not null check(Type IN(0,1)),
ItemContent varchar(50) not null,
Analysis varchar(200),
Del int not null default 0 check(Del IN(0,1))
);
```

试题表 t_question 的详细信息如表 9-3 所示。

表 9-3

名称	字段名	数据类型	备注
ID	ID	int(11)	主键，自增，每次增量为 1
题型	Type	int(11)	0 表示单选，1 表示多选，不能为空
题干	ItemContent	varchar(50)	不能为空
题典分析	Analysis	varchar(200)	可以为空
删除标识	Del	int(11)	0 表示正常，1 表示已删除，默认值为 0
分值	Point	int(11)	试题的分值，默认值为 0

（2）在当前数据库中创建一个选项表 t_option。

```
create table t_option
(
ID int not null primary key auto_increment,
QuestionID int not null,
Num int not null check(Num IN(0,1,2,3)),
OptionContent varchar(100) not null,
IsTrue int not null check(IsTrue IN(0,1)),
Del int not null default 0 check(Del IN(0,1)),
foreign key(QuestionID) references t_question(ID)
);
```

选项表 T_Option 的详细信息如表 9-4 所示。

表 9-4

名称	字段名	数据类型	备注
ID	ID	int(11)	主键，自增，每次增量为 1，不能为空
试题 ID	QuestionID	int(11)	外键，参照试题表，不能为空
选项号	Num	int(11)	0 表示 A，1 表示 B，2 表示 C，3 表示 D，不能为空
选项内容	OptionContent	varchar(100)	不能为空
标识是否是正确选项	IsTrue	int(11)	0 表示不是正确选项，1 表示是正确选项，不能为空
删除标识	Del	int(11)	0 表示正常，1 表示已删除，默认值为 0

（3）查看表结构，如图 9-5 所示。

```
DESC t_question;
```

```
mysql> desc t_question;
+-------------+--------------+------+-----+---------+----------------+
| Field       | Type         | Null | Key | Default | Extra          |
+-------------+--------------+------+-----+---------+----------------+
| ID          | int(11)      | NO   | PRI | NULL    | auto_increment |
| Type        | int(11)      | NO   |     | NULL    |                |
| ItemContent | varchar(50)  | NO   |     | NULL    |                |
| Analysis    | varchar(200) | YES  |     | NULL    |                |
| Del         | int(11)      | NO   |     | 0       |                |
+-------------+--------------+------+-----+---------+----------------+
5 rows in set (0.09 sec)
```

图 9-5

说明：
① Field：表示字段名。
② Type：表示字段数据类型。
③ NULL：表示该列是否可以存储 NULL 值。
④ Key：表示该列的键值。
⑤ Default：表示该列是否有默认值。
⑥ Extra：表示可以获取的与给定列有关的附加信息。
（4）复制试题表 t_question 结构，创建 t_questionCopy 表。
```
CREATE TABLE t_questionCopy (LIKE t_question);
```
查看 t_questionCopy 表结构，如图 9-6 所示。

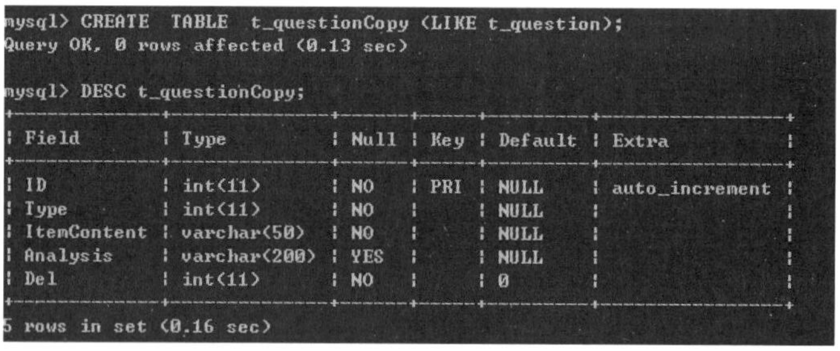

图 9-6

步骤 3：进行表的操作

（1）对试题表 t_question 进行表的修改操作。
① 向试题表增加"分数"属性，数据类型为 float，默认值为'0'。
```
alter table t_question add Score float(3,1) default '0';
```
② 修改字段类型，将 Score 的数据类型改为整数。
```
alter table t_question modify column Score int;
```
③ 修改字段名，将 Score 改为 Point。
```
alter table t_question change Score Point int default '0';
```
查看 t_question 表结构，如图 9-7 所示。

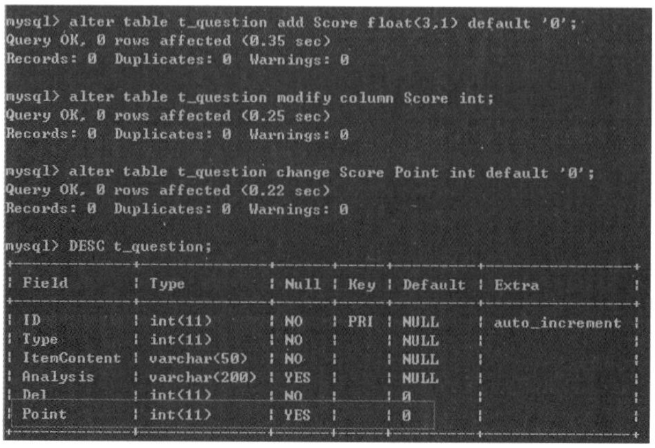

图 9-7

（2）修改表名，将 t_questionCopy 修改为 t_questionCopy1。
```
alter table t_questionCopy rename t_questionCopy1;
```
查看数据库中的所有表，如图9-8所示。

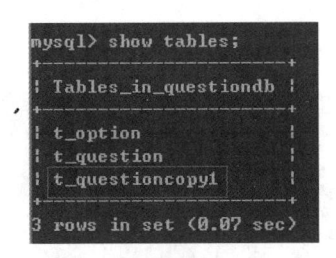

图 9-8

（3）对试题表 t_questionCopy1 进行表的删除操作。
① 删除 Point 字段。
```
alter table t_questionCopy1 drop Point;
```
② 删除试题表。
```
drop table t_questionCopy1 cascade;
```

步骤 4：创建视图

创建 t_question 表的视图，只显示 ID 列与 ItemContent 列。
```
create view v_question AS select ID,ItemContent from t_question;
```
视图：从一个或几个基本表（或视图）导出的表。视图本身没有数据，可以用于隐藏当前表的真实定义。

步骤 5：创建索引

为 t_question 表的 ID 列按降序创建一个唯一索引。
```
create unique INDEX QueID on t_question(ID) DESC;
```
索引：帮助 MySQL 高效获取数据的数据结构，用户可以根据需要在基本表上建立一个或多个索引，以提供多种存取路径，加快查找速度。

步骤 6：管理表数据

（1）向试题表 t_question 中插入一条记录。
```
insert into t_question(ID,Type,ItemContent,Analysis,Del)
values(1,0,'数据库系统的核心是','数据库知识',0);
```
查看表数据，如图9-9所示。

图 9-9

（2）向选项表 t_option 中插入 4 条记录。
```
insert into t_option values('1','1','0','数据模型','1','0'), ('2','1','1','数据库管理系统','0','0'), ('3','1','2','数据库','0','0'), ('4','1','3','数据库管理员','0','0');
```
查看表数据，如图 9-10 所示。

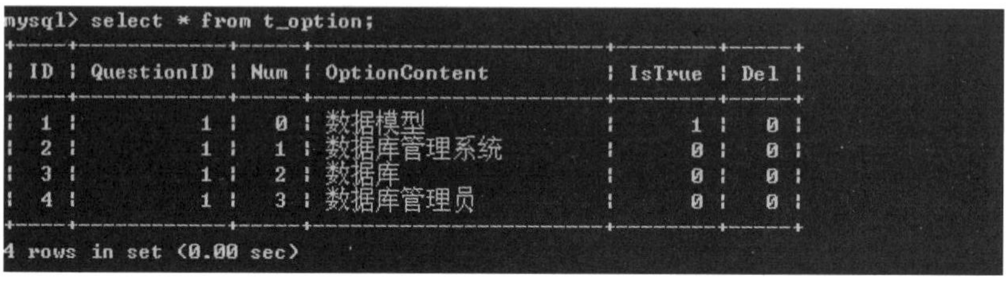

图 9-10

（3）删除 t_question 表中 ID 为 2 的行。
```
delete from t_question where ID=2;
```
查看表数据，如图 9-11 所示。

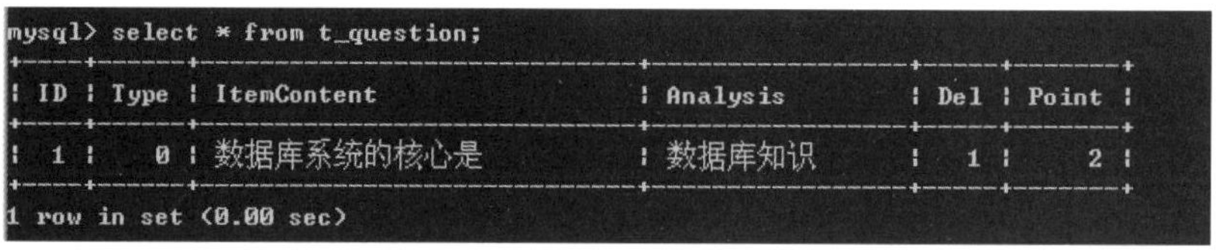

图 9-11

步骤 7：查询表数据

（1）查询 t_question 表的 ID 与 ItemContent 列。
```
select ID,ItemContent from t_question;
```
执行结果如图 9-12 所示。

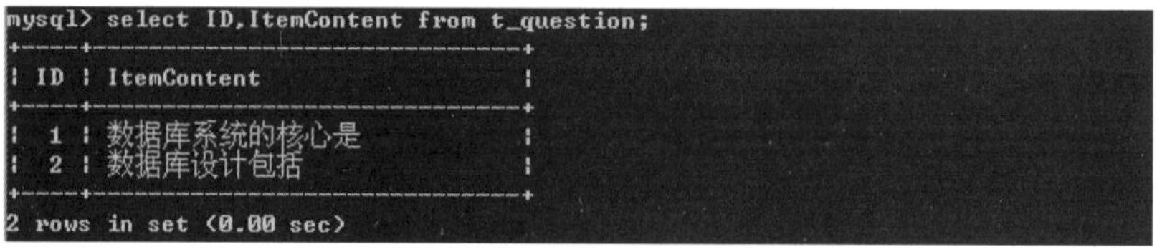

图 9-12

（2）查询 t_question 表中的所有列。
查询全部列可以通过以下两种方法实现。
① 在 SELECT 关键字后面列出所有属性列名。
② 在 SELECT 关键字后使用星号（*），属性列名的顺序与其在基本表中的顺序相同。
```
select * from t_question;
select ID,Type,ItemContent,Analysis,Del,Point from t_question;
```
执行结果如图 9-13 所示。

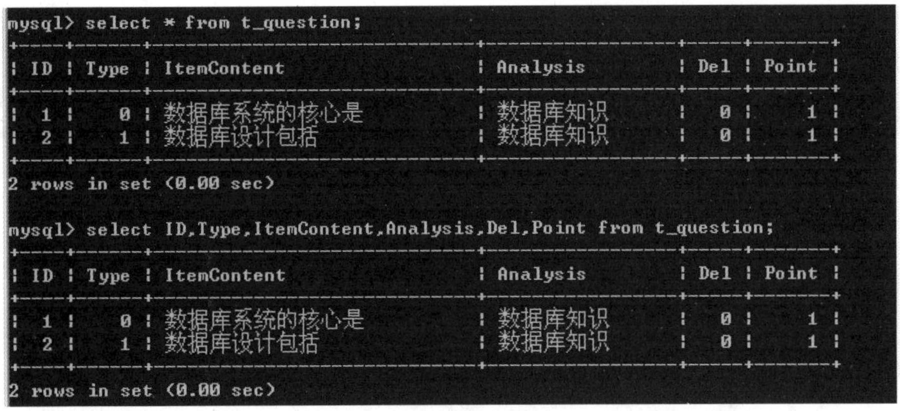

```
mysql> select * from t_question;
+----+------+-----------------------+--------------+-----+-------+
| ID | Type | ItemContent           | Analysis     | Del | Point |
+----+------+-----------------------+--------------+-----+-------+
|  1 |    0 | 数据库系统的核心是    | 数据库知识   |   0 |     1 |
|  2 |    1 | 数据库设计包括        | 数据库知识   |   0 |     1 |
+----+------+-----------------------+--------------+-----+-------+
2 rows in set (0.00 sec)

mysql> select ID,Type,ItemContent,Analysis,Del,Point from t_question;
+----+------+-----------------------+--------------+-----+-------+
| ID | Type | ItemContent           | Analysis     | Del | Point |
+----+------+-----------------------+--------------+-----+-------+
|  1 |    0 | 数据库系统的核心是    | 数据库知识   |   0 |     1 |
|  2 |    1 | 数据库设计包括        | 数据库知识   |   0 |     1 |
+----+------+-----------------------+--------------+-----+-------+
2 rows in set (0.00 sec)
```

图 9-13

（3）查询满足条件的记录。

查询 t_question 表中 ID 小于 2 或大于 3 的所有记录。

`select * from t_question where ID<2 OR ID>3;`

执行结果如图 9-14 所示。

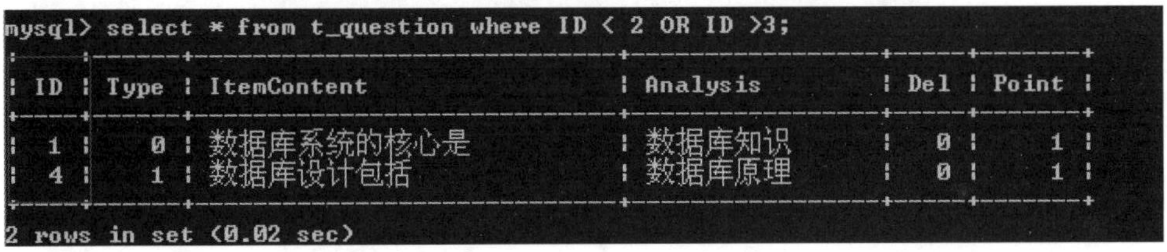

```
mysql> select * from t_question where ID < 2 OR ID >3;
+----+------+-----------------------+--------------+-----+-------+
| ID | Type | ItemContent           | Analysis     | Del | Point |
+----+------+-----------------------+--------------+-----+-------+
|  1 |    0 | 数据库系统的核心是    | 数据库知识   |   0 |     1 |
|  4 |    1 | 数据库设计包括        | 数据库原理   |   0 |     1 |
+----+------+-----------------------+--------------+-----+-------+
2 rows in set (0.02 sec)
```

图 9-14

数据库谓词如表 9-5 所示。

表 9-5

查询条件	谓词
比较	=、>、<、>=、<=、!=、!>、!<、<>、NOT+上述比较符
确定范围	BETWEEN AND、NOT BETWEEN AND
确定集合	IN、NOT IN
字符匹配	LIKE、NOT LIKE
空值	IS NULL、IS NOT NULL
多重条件（逻辑运算）	AND、OR、NOT

步骤 8：导入和导出数据库脚本

（1）导出整个数据库。

`mysqldump 指令：mysqldump -u [username] -p [-opt] [databasename]>[filepath] //（其中[-opt]是参数设置,可选）`

将数据库 questiondb 导出到 "D:/mysql_study" 文件夹中，如图 9-15 所示。

图 9-15

查看导出文件，如图 9-16 所示。

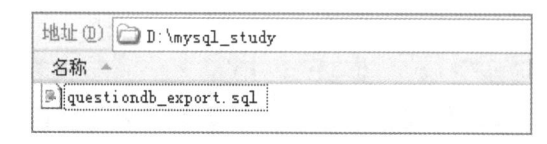

图 9-16

（2）导入 SQL 脚本。进入 questiondb 数据库并导入"D:/mysql_study/questiondb.sql"位置的脚本，如图 9-17 所示。

```
source D:/mysql_study/questiondb.sql;
```

注意：脚本路径使用"/"隔开，而不是反斜杠"\"。

图 9-17

步骤 9：事务控制

1. 开启事务

开启事务可以使用如下语句：
```
START TRANSACTION;
```
或
```
BEGIN WORK;
```
两条语句都可以用来开启事务，但是 START TRANSACTION 更为常用。

注意：事务不可以嵌套，当第二个事务开始时会自动提交第一个事务。

2. 事务回滚

开启事务，并将 t_question 表中 ID 为 1 的记录的 Del 字段值修改为 1，之后进行事务回滚，查看修改结果。

```
start transaction;
update t_question set Del=1 where ID=1;
rollback;
select * from t_question;
```

执行结果如图 9-18 所示。

图 9-18

3. 事务确认

开启事务,设置 t_question 表与 t_option 表同一试题的 Del 字段值为 1,然后确认提交,查看修改结果。

```
start transaction;
update t_question set Del=1 where ID=1;
update t_option set Del=1 where QuestionID=1;
commit;
select * from t_question;
```

执行结果如图 9-19 所示。

图 9-19

4. 自动提交

MySQL 默认的执行方式是自动提交,也就是 SQL 语句执行完毕后会自动提交工作。可以使用 SET 语句设置 MySQL 是否自动提交。

开启自动提交:
```
SET AUTOCOMMINT=1;
```
关闭自动提交:
```
SET AUTOCOMMINT=0;
```
查看自动提交当前的状态:
```
show variables like 'autocommit';
```
显示结果如图 9-20 所示。

```
mysql> show variables like 'autocommit';
+---------------+-------+
| Variable_name | Value |
+---------------+-------+
| autocommit    | ON    |
+---------------+-------+
1 row in set (0.00 sec)
```

图 9-20

Value 为 ON，则表示当前状态为开启自动提交。

步骤 10：创建触发器

（1）为了使选项表与相应试题表的删除标志 Del 同步更新，创建如下触发器。

```
create trigger shanchu after update on t_question for each row
update t_option set Del=(select Del from t_question where t_option.QuestionID=
t_question.ID);
```

（2）有了触发器，就可以通过 UPDATE 语句来激活触发器。

更新试题表，设置 ID 为 1 的记录的 Del 值为 1。

```
update t_question set Del=1 where ID=1;
```

（3）执行结果如下。

① 创建触发器，如图 9-21 所示。

```
mysql> create trigger shanchu after update on t_question for each row
    -> update t_option set Del=(select Del from t_question where t_option.Questi
onID = t_question.ID);
Query OK, 0 rows affected (0.03 sec)
```

图 9-21

② 激活触发器，如图 9-22 所示。

```
mysql> update t_question set Del = 1 where ID = 1;
Query OK, 1 row affected (0.17 sec)
Rows matched: 1  Changed: 1  Warnings: 0
```

图 9-22

③ 查看更新前的结果，如图 9-23 所示。

```
mysql> select * from t_question;
+----+------+--------------------+----------+-----+-------+
| ID | Type | ItemContent        | Analysis | Del | Point |
+----+------+--------------------+----------+-----+-------+
|  1 |    0 | 数据库系统的核心是 | 数据库知识 |   1 |     1 |
|  2 |    1 | 数据库设计包括     | 数据库知识 |   0 |     1 |
+----+------+--------------------+----------+-----+-------+
2 rows in set (0.00 sec)
```

图 9-23

④ 查看更新后的结果，如图 9-24 所示。

```
mysql> select * from t_option;
+----+------------+-----+----------------+--------+-----+
| ID | QuestionID | Num | OptionContent  | IsTrue | Del |
+----+------------+-----+----------------+--------+-----+
|  1 |          1 |   0 | 数据模型       |      1 |   1 |
|  2 |          1 |   1 | 数据库管理系统 |      0 |   1 |
|  3 |          1 |   2 | 数据库         |      0 |   1 |
|  4 |          1 |   3 | 数据库管理员   |      0 |   1 |
```

图 9-24

（4）删除触发器。

```
DROP TRIGGER [IF EXISTS ] shanchu;
```

执行结果如图 9-25 所示。

```
mysql> DROP TRIGGER if exists shanchu ;
Query OK, 0 rows affected (0.00 sec)
```

图 9-25

步骤 11：创建存储过程

创建一个存储过程，新增一道试题，向试题表和选项表分别添加一条记录。

```
delimiter $$
create procedure xinzeng()
BEGIN
insert into t_question(Type,ItemContent,Analysis,Del) values (0,'新增试题','存储过程',0);
insert into t_option(QuestionID,Num,OptionContent,IsTrue) values ('1','0','新增选项','1');
END
$$
delimiter ;
```

（1）调用存储过程。

MySQL 可以使用 CALL 来调用存储过程。

```
CALL xinzeng();
```

（2）查看存储过程状态。

```
SHOW PROCEDURE STATUS \G
```

效果如图 9-26 所示。

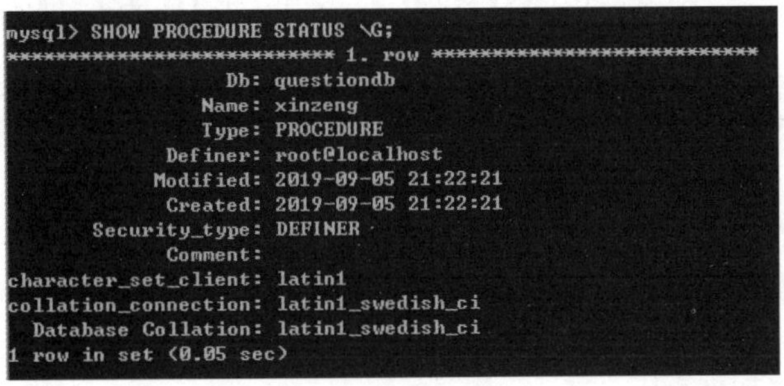

```
mysql> SHOW PROCEDURE STATUS \G;
*************************** 1. row ***************************
                  Db: questiondb
                Name: xinzeng
                Type: PROCEDURE
             Definer: root@localhost
            Modified: 2019-09-05 21:22:21
             Created: 2019-09-05 21:22:21
       Security_type: DEFINER
             Comment:
character_set_client: latin1
collation_connection: latin1_swedish_ci
  Database Collation: latin1_swedish_ci
1 row in set (0.05 sec)
```

图 9-26

（3）删除存储过程。

```
DROP PROCEDURE [IF EXISTS ] xinzeng;
```

第 10 章

数据库编程（JDBC）：学生成绩管理

10.1 实验目标

（1）掌握 Java 的基础语法、编码规范。
（2）掌握 MySQL 的基本操作方法。
（3）掌握 JDBC 操作 MySQL 的方法。
（4）掌握 JDBC 执行 SQL 语句的方法。
（5）掌握 JDBC 采用预处理的方式执行 SQL 语句的方法。
（6）掌握 JDBC 绑定预处理参数的操作方法。
（7）掌握 JDBC 数据库查询记录集的操作方法。
（8）掌握会话 Session 的操作方法。
（9）综合应用 Java 数据库操作技术，编写"学生成绩管理"程序。

10.2 实验任务

编写一个学生成绩管理程序。

1. 用户登录

进入登录页面，输入用户名和密码，如图 10-1 所示，若验证成功则进入学生成绩管理系统的主页——成绩列表页面。

学生成绩管理系统

用户名：	user
密码：	••••
登录	

图 10-1

2．成绩列表页面

页头显示"学生成绩管理系统",页面内容显示所有的成绩列表。每条记录显示 id 值、姓名、年龄和成绩,以及"修改"与"删除"按钮。在列表的最后一行显示"添加"按钮,如图 10-2 所示。

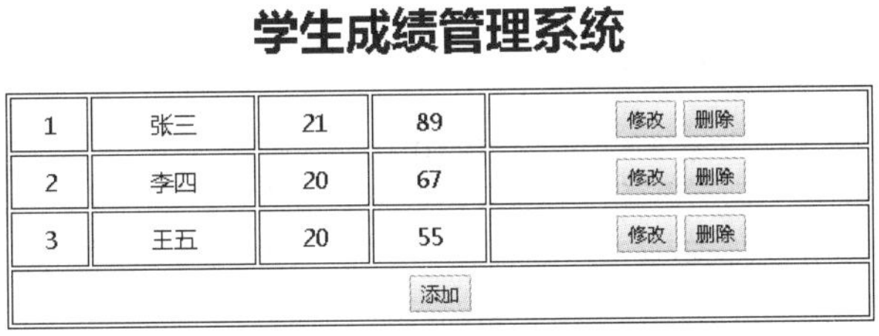

学生成绩管理系统

1	张三	21	89	修改	删除
2	李四	20	67	修改	删除
3	王五	20	55	修改	删除
添加					

图 10-2

（1）单击"删除"按钮,删除当前记录,并更新列表页面。
（2）单击"修改"按钮,进入成绩修改页面。
（3）单击"添加"按钮,进入成绩添加页面。

3．成绩修改页面

成绩修改页面显示姓名、年龄和成绩的输入框,初始时显示当前记录的数据,在表格最后一行显示"修改"按钮,如图 10-3 所示。修改完成后单击"修改"按钮返回成绩列表页面。

学生成绩管理系统

姓名：	王五
年龄：	20
成绩：	65
修改	

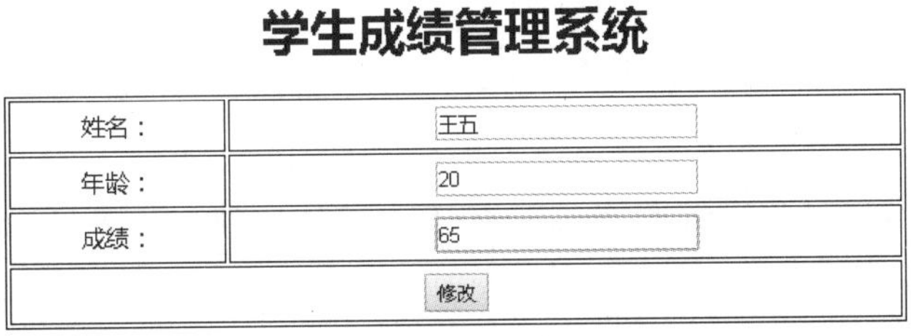

图 10-3

4．成绩添加页面

成绩添加页面显示姓名、年龄和成绩的输入框,在表格最后一行显示"添加"按钮,如图 10-4 所示。添加完成后单击"添加"按钮返回成绩列表页面。

学生成绩管理系统

姓名：	乔安
年龄：	21
成绩：	89
添加	

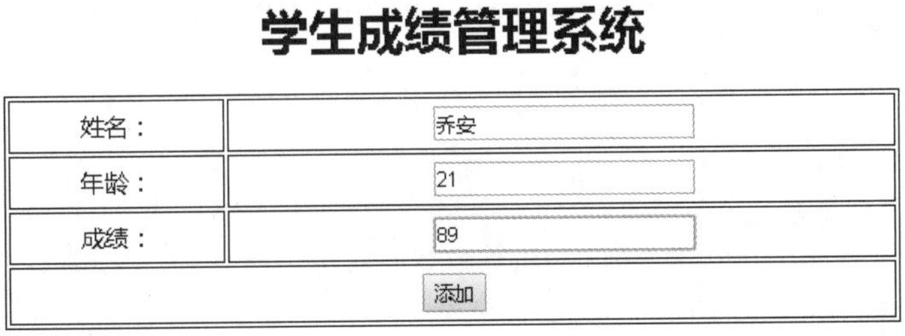

图 10-4

10.3 设计思路

1．工程设计

创建 Dynamic Web Project 项目——StuManage。

2．文件设计

项目文件如表 10-1 所示。

表 10-1

类型	文件／类	说明
JSP 文件	index.jsp	学生成绩管理首页
	login.jsp	登录页面
	insert.jsp	成绩添加页面
	update.jsp	成绩修改页面
Java 类	com.stu.servlet.InsertServlet	输入成绩
	com.stu.servlet.RemoveServlet	删除成绩
	com.stu.servlet.SelectServlet	查询单个成绩
	com.stu.servlet.UpdateServlet	更改成绩
	com.stu.servlet.UserServlet	用户登录
	com.stu.servlet.Conn	创建数据库连接
	com.stu.model.Result	学生成绩记录类
	com.stu.model.User	用户类
css 文件	css/style.css	页面样式

3．数据库设计

创建学生成绩管理数据库脚本 db.sql，代码如下，创建 stu_result 数据库，创建 result 和 user 表。

```
DROP DATABASE IF EXISTS stu_result;
CREATE DATABASE stu_result;
USE stu_result;

SET NAMES utf8mb4;
SET FOREIGN_KEY_CHECKS=0;

-- ----------------------------
-- Table structure for 'result'
-- ----------------------------
DROP TABLE IF EXISTS 'result';
CREATE TABLE 'result'(
  'id' int(11) NOT NULL AUTO_INCREMENT,
  'name' varchar(64) NOT NULL,
  'age' int(11) NOT NULL,
  'result' varchar(255) DEFAULT NULL,
  PRIMARY KEY ('id')
) ENGINE=InnoDB AUTO_INCREMENT=6 DEFAULT CHARSET=utf8;
```

```
-- ---------------------------
-- Records of result
-- ---------------------------
INSERT INTO 'result' VALUES ('1', '张三', '21', '89');
INSERT INTO 'result' VALUES ('2', '李四', '20', '67');
INSERT INTO 'result' VALUES ('3', '王五', '20', '55');
-- ---------------------------
-- Table structure for 'user'
-- ---------------------------
DROP TABLE IF EXISTS 'user';
CREATE TABLE 'user'(
  'id' int(11) NOT NULL AUTO_INCREMENT,
  'account' varchar(32) NOT NULL,
  'password' varchar(32) NOT NULL,
  PRIMARY KEY ('id')
) ENGINE=InnoDB AUTO_INCREMENT=2 DEFAULT CHARSET=utf8;

-- ---------------------------
-- Records of user
-- ---------------------------
INSERT INTO 'user' VALUES ('1', 'user', 'user');
```

成绩表包括 4 个字段：主键、学生姓名、学生年龄和学生成绩，如表 10-2 所示。

表 10-2

字段	说明
Id	主键
Name	学生姓名
age	学生年龄
result	学生成绩

用户表包括 3 个字段：主键、用户名和密码，如表 10-3 所示。

表 10-3

字段	说明
id	主键
account	用户名
password	密码

4．实现设计

用户登录系统，使用表单进行请求，请求 login.jsp 文件进行处理，若验证成功则跳转到学生成绩管理首页 index.jsp。

学生成绩管理首页显示成绩列表，可以通过页面的"添加"和"修改"按钮进入成绩添加页面（insert.jsp）和成绩修改页面（update.jsp）；可以通过单击"删除"按钮进行删除操作，删除操作处理类为 RemoveServlet；添加、修改、删除操作处理完成后跳转回学生成绩管理首页，刷新成绩列表。

在成绩添加页面和成绩修改页面使用 AJAX 进行静态请求，使用 JSON 进行数据交互，并使用 JavaScript 静态更新页面数据。

定义数据库连接类 Conn。设置数据库服务器地址、用户名、密码和数据库名。创建 Connection 对象，连接数据库。

5. 页面设计

1）成绩列表页面

页面：index.jsp。

页面使用<table>标签显示成绩列表，共 4 列，最后一列为操作按钮。

将表格边框的 border 值设置为 1。

表格最后一行，显示<button>"添加"按钮。

页面加载时，调用 ServletServlet 类文件查询出当前数据库 result 表中的所有记录，使用 for 显示所有的记录。

单击"修改"按钮，跳转到成绩修改页面，将当前记录的 id 值作为参数传到成绩修改页面。

单击"删除"按钮，删除当前记录，将当前记录的 id 值作为参数传到 RemoveServlet 处理函数，使用 JDBC 类实现 MySQL 数据库操作，删除记录。

2）成绩添加页面

页面：insert.jsp。

页面使用<table>标签编写表单，共 2 列，第 1 列为字段标签，第 2 列为输入框，分别显示姓名、年龄和成绩。

将表格边框的 border 值设置为 1。

表格最后一行显示<button>"添加"按钮。

通过 JavaScript DOM 操作提交表单请求/Insert，使用 InsertServlet 类实现 MySQL 数据库操作，添加记录。

3）成绩修改页面

页面：update.jsp。

页面使用<table>标签编写表单，共 2 列，第 1 列为字段标签，第 2 列为输入框，分别显示姓名、年龄和成绩。

将表格边框的 border 值设置为 1。

表格最后一行显示<button>"修改"按钮。

在页面开始通过 id 查询记录，并使用<%=%>脚本显示当前记录原有的值。

表单提交请求/Update，使用 UpdateServlet 类实现 MySQL 数据库操作，更新记录。

通过 JavaScript DOM 操作请求 Java 页面，Java 页面使用 PrintWriter 类返回结果。

4）登录页面

页面：login.jsp。

页面使用<table>标签编写表单，共 2 列，第 1 列为字段标签，第 2 列为输入框，分别显示用户名和密码。

将表格边框的 border 值设置为 1。

表格最后一行显示<button>"登录"按钮。

表单提交请求/Login，使用 UserServlet 类实现 MySQL 数据库操作，查询用户信息。

10.4 实验实施（跟我做）

步骤1：创建项目和文件

（1）创建项目：项目名为 stu_result。
（2）创建文件，如图 10-5 所示。

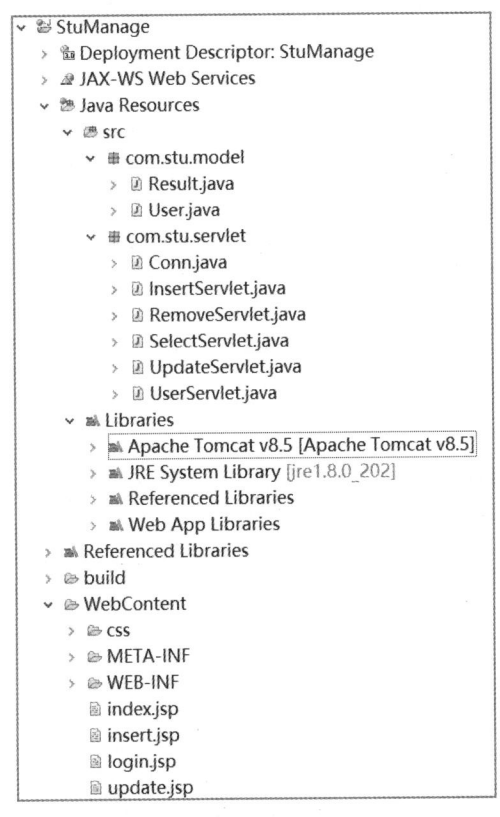

```
v ⯎ StuManage
   > ⯎ Deployment Descriptor: StuManage
   > ⯎ JAX-WS Web Services
   v ⯎ Java Resources
      v ⯎ src
         v ⯎ com.stu.model
            > ⯎ Result.java
            > ⯎ User.java
         v ⯎ com.stu.servlet
            > ⯎ Conn.java
            > ⯎ InsertServlet.java
            > ⯎ RemoveServlet.java
            > ⯎ SelectServlet.java
            > ⯎ UpdateServlet.java
            > ⯎ UserServlet.java
      v ⯎ Libraries
         > ⯎ Apache Tomcat v8.5 [Apache Tomcat v8.5]
         > ⯎ JRE System Library [jre1.8.0_202]
         > ⯎ Referenced Libraries
         > ⯎ Web App Libraries
   > ⯎ Referenced Libraries
   > ⯎ build
   v ⯎ WebContent
      > ⯎ css
      > ⯎ META-INF
      > ⯎ WEB-INF
         ⯎ index.jsp
         ⯎ insert.jsp
         ⯎ login.jsp
         ⯎ update.jsp
```

图 10-5

index.jsp：学生成绩管理首页。
login.jsp：登录页面。
insert.Java：成绩添加页面。
update.Java：成绩修改页面。
com.stu.servlet.Conn：创建数据库连接类。
com.stu.servlet.InsertServlet：输入成绩。
com.stu.servlet.RemoveServlet：删除成绩。
com.stu.servlet.SelectServlet：查询单个成绩。
com.stu.servlet.UserServlet：用户登录。
com.stu.servlet.UpdateServlet：更改成绩。
com.stu.model.User：用户类（与 user 表对应）
ocm.stu.model.Result：学生成绩记录类（与 result 表对应）
css/style.css：页面样式。

步骤 2：创建数据库

启动 MySQL 命令行，输入"mysql -uroot -p"命令登录 MySQL 数据库，使用 source 命令导入数据库脚本 db.sql，脚本内容参考设计思路中的"创建学生成绩管理数据库脚本"。

步骤 3：制作 CSS 样式

在 style.css 文件中编写<h1>、<table>、<th>、<td>和<button>标签的样式，制作 CSS 样式，代码如下：

```css
h1{
    text-align: center;
}
table{
    width:600px;
    border:1px solid #000000;
    text-align: center;
    margin: 0 auto;
}
th,td{
    padding: 5px;
    border:1px solid #000000;
}
.button{
    width: 280px;
    margin: 0 2px;
}
```

步骤 4：创建登录页面

（1）创建 login.jsp 文件，编写代码，实现登录页面。

使用 input 文本控件接收用户名和密码登录系统。

使用 form 表单提交登录信息，action 的值为 Login，方法为 POST。

在登录按钮之后的<span>标签中，显示异常消息。

代码如下：

```jsp
<%@ page language="java" contentType="text/html; charset=UTF-8"
    pageEncoding="UTF-8"%>
<!DOCTYPE html>
<html>
<head>
<meta charset="UTF-8">
<link rel="stylesheet" type="text/css" href="css/style.css"/>
    <title>学生成绩管理系统</title>
</head>
<body>
<h1>学生成绩管理系统</h1>
    <form action="Login" method="post">
```

```html
        <table>
            <tr>
                <td>用户名：</td>
                <td><input type="text" name="account"></td>
            </tr>
            <tr>
                <td>密码：</td>
                <td><input type="password" name="password"></td>
            </tr>
            <tr>
                <td><button type="submit">登录</button></td>
                <td>
                <span>
                <%
                if(request.getAttribute("message")!=null){
                   out.print(request.getAttribute("message"));
                }
                %>
                </span>
                </td>
            </tr>
        </table>
    </form>
</body>
</html>
```

(2) 创建 com.stu.servlet.Conn 包，在包中创建 Conn 类。

```java
package com.stu.servlet;

import java.sql.Connection;
import java.sql.DriverManager;
import java.sql.SQLException;

//MySQL 数据库连接工具类
public class Conn {
    private static String server="127.0.0.1";        //数据库服务器地址
    private static String username="root";           //登录用户名
    private static String password="123456";         //密码
    private static String dbname="stu_result";       //数据库名称

    //返回连接对象
    public static Connection getConnection() throws ClassNotFoundException, SQLException {
        //注册驱动（MySQL 6 版本以下使用 com.mysql.jdbc.Driver 类）
        Class.forName("com.mysql.cj.jdbc.Driver");
        //拼接连接字符串
        String connectString=String.format("jdbc:mysql://%s:3306/%s?useUnicode=true&characterEncoding=UTF-8", server, dbname);
```

```
        //获得数据库连接对象
        return DriverManager.getConnection(connectString, username, password);
    }
}
```

将 jar 包 mysql-connector-java-8.0.23.jar 复制到"WebContent/WEB-INF/lib"文件夹中，右击要复制的包，在弹出的快捷菜单中选择"Build Path"→"Add to Build Path"选项，将 jar 包添加到工程中，如图 10-6 所示。

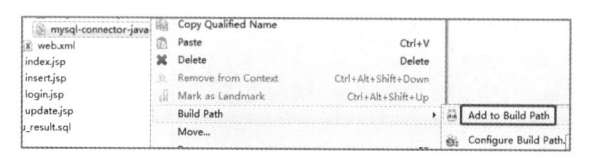

图 10-6

在工程中的"Referenced Libraries"中可以看到添加的 jar 包，如图 10-7 所示

图 10-7

（3）在 com.stu.servlet 包中创建 UserServlet 类，在该类中编写 Java 代码，对用户的账号和密码进行验证。

使用 import 导入 Conn 类。

从 request 中获取账号和密码信息。

编写 SQL 语句，使用账号和密码查询用户，调用 executeQuery( )函数。

若查询到了记录，则表示登录成功，启动会话，在 Session 中存储账号信息。

使用 sendRedirect( )函数跳转到 index.jsp 页面。

若查询不到记录，则跳转到 login.jsp 页面，并显示提示信息"用户名或密码不一致"。

```java
package com.stu.servlet;

import java.io.IOException;
import java.sql.Connection;
import java.sql.ResultSet;
import java.sql.SQLException;
import java.sql.Statement;

import javax.servlet.ServletException;
import javax.servlet.annotation.WebServlet;
import javax.servlet.http.HttpServlet;
import javax.servlet.http.HttpServletRequest;
import javax.servlet.http.HttpServletResponse;
import javax.servlet.http.HttpSession;

/**
 * Servlet implementation class UserServlet
 */
```

```java
@WebServlet("/Login")
public class UserServlet extends HttpServlet{
    private static final long serialVersionUID=1L;

    /**
     * @see HttpServlet#HttpServlet()
     */
    public UserServlet(){
        super();
        // TODO Auto-generated constructor stub
    }

    /**
     * @see HttpServlet#doPost(HttpServletRequest request, HttpServletResponse
     *      response)
     */
    protected void doPost(HttpServletRequest request, HttpServletResponse response)
            throws ServletException, IOException{
        //获得账号和密码
        String account=request.getParameter("account");
        String password=request.getParameter("password");
        //获得Session对象
        HttpSession session=request.getSession();
        //执行查询操作
        Connection conn=null;
        Statement stat=null;
        ResultSet rs=null;
        try{
            //获得连接对象
            conn=Conn.getConnection();
            //拼接SQL语句
            String sql=String.format("select id from user where account='%s' and password='%s'", account, password);
            //创建Statement对象
            stat=conn.createStatement();
            //执行
            rs=stat.executeQuery(sql);
            if(rs.next()){
                session.setAttribute("user.account", account);
                response.sendRedirect(request.getContextPath() +"/index.jsp");
            } else{
                request.setAttribute("message", "用户或密码不一致");
                request.getRequestDispatcher("login.jsp").forward(request, response);
            }
        } catch(ClassNotFoundException e){
```

```
          e.printStackTrace();
        } catch(SQLException e){
          e.printStackTrace();
        } finally{
          if(rs!=null){
            try{
              rs.close();
            } catch(SQLException e){
              e.printStackTrace();
            }
          }
          if(stat!=null){
            try{
              stat.close();
            }catch(SQLException e){
              //TODO Auto-generated catch block
              e.printStackTrace();
            }
          }
          if(conn!=null){
            try{
              conn.close();
            } catch(SQLException e){
              //TODO Auto-generated catch block
              e.printStackTrace();
            }
          }
        }
      }
    }
```

步骤 5：编写数据库操作 Java 代码

（1）在 com.stu.servlet 包中创建 InsertServlet 类，在 InsertServlet::doPost( )方法中编写输入成绩代码。

包含 Conn 类。

使用 request.getParameter 方法获得 name、age、result。

创建预处理语句：使用 conn.prepareStatement(psql);创建预处理语句。

绑定数据：使用 ps.setString(1,name)、ps.setInt(2,age)对预处理语句进行数据绑定。

执行预处理语句：ps.executeUpdate( )。

关闭连接：在 finally 语句中关闭 ps 和 conn。

输入的成绩代码如下：

```
package com.stu.servlet;

import java.io.IOException;
import java.sql.Connection;
```

```java
import java.sql.PreparedStatement;
import java.sql.SQLException;

import javax.servlet.ServletException;
import javax.servlet.annotation.WebServlet;
import javax.servlet.http.HttpServlet;
import javax.servlet.http.HttpServletRequest;
import javax.servlet.http.HttpServletResponse;

/**
 * Servlet implementation class InsertServlet
 */
@WebServlet("/Insert")
public class InsertServlet extends HttpServlet{
    private static final long serialVersionUID=1L;

    /**
     * @see HttpServlet#HttpServlet()
     */
    public InsertServlet(){
        super();
        //TODO Auto-generated constructor stub
    }

    /**
     * @see HttpServlet#doPost(HttpServletRequest request, HttpServletResponse
     *      response)
     */
    protected void doPost(HttpServletRequest request, HttpServletResponse response)
        throws ServletException, IOException{
        //获得学生成绩信息
        String name=request.getParameter("name");
        int age=Integer.parseInt(request.getParameter("age"));
        String result=request.getParameter("result");
        //添加学生成绩信息
        Connection conn=null;
        PreparedStatement ps=null;
        try{
            conn=Conn.getConnection();
            String psql="insert into result(name, age, result) values(?, ?, ?)";
            ps=conn.prepareStatement(psql);
            ps.setString(1, name);
            ps.setInt(2, age);
            ps.setString(3, result);
            ps.executeUpdate();
        } catch(ClassNotFoundException e){
```

```
        e.printStackTrace();
      } catch(SQLException e){
        e.printStackTrace();
      } finally{
        if(ps!=null){
          try{
            ps.close();
          } catch(SQLException e){
            e.printStackTrace();
          }
        }
        if(conn!=null){
          try{
            conn.close();
          } catch(SQLException e){
            e.printStackTrace();
          }
        }
      }
    }
}
```

（2）在 RemoveServlet 类中编写删除成绩代码。

删除成绩的代码如下：

```
package com.stu.servlet;

import java.io.IOException;
import java.sql.Connection;
import java.sql.SQLException;
import java.sql.Statement;

import javax.servlet.ServletException;
import javax.servlet.annotation.WebServlet;
import javax.servlet.http.HttpServlet;
import javax.servlet.http.HttpServletRequest;
import javax.servlet.http.HttpServletResponse;

/**
 * Servlet implementation class RemoveServlet
 */
@WebServlet("/Remove")
public class RemoveServlet extends HttpServlet{
  private static final long serialVersionUID=1L;

  /**
   * @see HttpServlet#HttpServlet()
   */
  public RemoveServlet(){
```

```java
        super();
        //TODO Auto-generated constructor stub
    }

    /**
     * @see HttpServlet#doGet(HttpServletRequest request, HttpServletResponse
     *      response)
     */
    protected void doGet(HttpServletRequest request, HttpServletResponse response)
        throws ServletException, IOException{
        //获得要删除的id
        int id=Integer.parseInt(request.getParameter("id"));
        //删除记录
        Connection conn=null;
        Statement stat=null;
        try{
            conn=Conn.getConnection();
            String sql=String.format("delete from result where id=%d", id);
            stat=conn.createStatement();
            stat.execute(sql);
            response.sendRedirect("index.jsp");
        } catch(ClassNotFoundException e){
            e.printStackTrace();
        } catch(SQLException e){
            e.printStackTrace();
        } finally{
            if(stat!=null){
                try{
                    stat.close();
                } catch(SQLException e){
                    e.printStackTrace();
                }
            }
            if(conn!=null){
                try{
                    conn.close();
                } catch(SQLException e){
                    e.printStackTrace();
                }
            }
        }
    }
}
```

(3) 编写 Result 类。

```
package com.stu.model;
```

```java
import com.alibaba.fastjson.annotation.JSONField;

public class Result{
  //设置该属性转为JSON字符串时的key值
  @JSONField(name="id")
  private int id;              //标识符
  @JSONField(name="name")
  private String name;         //学生姓名
  @JSONField(name="age")
  private int age;             //年龄
  @JSONField(name="result")
  private String result;       //成绩

  public Result(){

  }

  public Result(int id, String name, int age, String result){
    super();
    this.id=id;
    this.name=name;
    this.age=age;
    this.result=result;
  }

  public int getId(){
    return id;
  }

  public void setId(int id){
    this.id=id;
  }

  public String getName(){
    return name;
  }

  public void setName(String name){
    this.name=name;
  }

  public int getAge(){
    return age;
  }

  public void setAge(int age){
    this.age=age;
```

```
  }
  public String getResult(){
    return result;
  }

  public void setResult(String result){
    this.result=result;
  }
}
```

（4）在 SelectServlet 中编写查询单个成绩的代码。

查询单个成绩的代码如下：

```
package com.stu.servlet;

import java.io.IOException;
import java.io.PrintWriter;
import java.sql.Connection;
import java.sql.ResultSet;
import java.sql.SQLException;
import java.sql.Statement;

import javax.servlet.ServletException;
import javax.servlet.annotation.WebServlet;
import javax.servlet.http.HttpServlet;
import javax.servlet.http.HttpServletRequest;
import javax.servlet.http.HttpServletResponse;

import com.alibaba.fastjson.JSON;
import com.stu.model.Result;

/**
 * Servlet implementation class SelectServlet
 */
@WebServlet("/Select")
public class SelectServlet extends HttpServlet{
  private static final long serialVersionUID=1L;

  /**
   * @see HttpServlet#HttpServlet()
   */
  public SelectServlet(){
    super();
    //TODO Auto-generated constructor stub
  }

  /**
   * @see HttpServlet#doGet(HttpServletRequest request, HttpServletResponse
```

```java
 *        response)
 */
protected void doGet(HttpServletRequest request, HttpServletResponse response)
    throws ServletException, IOException{
    //获得查询的记录id值
    int id=Integer.parseInt(request.getParameter("id"));
    //查询记录
    Connection conn=null;
    Statement stat=null;
    ResultSet rs=null;
    try{
      conn=Conn.getConnection();
      String sql=String.format("select * from result where id=%d", id);
      stat=conn.createStatement();
      Result result=new Result();
      rs=stat.executeQuery(sql);
      if(rs.next()){
        result.setId(rs.getInt("id"));
        result.setName(rs.getString("name"));
        result.setAge(rs.getInt("age"));
        result.setResult(rs.getString("result"));
      }
      //返回JSON字符串
      String json=JSON.toJSONString(result);
      response.setCharacterEncoding("UTF-8");
      response.setContentType("application/json; charset=utf-8");
      PrintWriter writer=response.getWriter();
      writer.append(json);
    } catch (ClassNotFoundException e){
      e.printStackTrace();
    } catch (SQLException e){
      e.printStackTrace();
    } finally{
      if(rs!=null){
        try{
          rs.close();
        } catch(SQLException e){
          e.printStackTrace();
        }
      }
      if(stat!=null){
        try{
          stat.close();
        } catch(SQLException e){
          e.printStackTrace();
        }
```

```
        }
      if(conn!=null){
        try{
          conn.close();
        } catch (SQLException e){
          e.printStackTrace();
        }
      }
    }
  }
}
```

（5）在 UpdateServlet 类中编写更改成绩代码。
更改成绩的代码如下：

```java
package com.stu.servlet;

import java.io.IOException;
import java.io.PrintWriter;
import java.sql.Connection;
import java.sql.SQLException;
import java.sql.Statement;

import javax.servlet.ServletException;
import javax.servlet.annotation.WebServlet;
import javax.servlet.http.HttpServlet;
import javax.servlet.http.HttpServletRequest;
import javax.servlet.http.HttpServletResponse;

/**
 * Servlet implementation class UpdateServlet
 */
@WebServlet("/Update")
public class UpdateServlet extends HttpServlet{
  private static final long serialVersionUID=1L;

  /**
   * @see HttpServlet#HttpServlet()
   */
  public UpdateServlet(){
    super();
    //TODO Auto-generated constructor stub
  }

  /**
   * @see HttpServlet#doPost(HttpServletRequest request, HttpServletResponse
   *      response)
   */
  protected void doPost(HttpServletRequest request, HttpServletResponse
```

```java
response)
        throws ServletException, IOException{
    //获得学生成绩信息
    int id=Integer.parseInt(request.getParameter("id"));
    String name=request.getParameter("name");
    int age=Integer.parseInt(request.getParameter("age"));
    String result=request.getParameter("result");
    //设置返回编码格式
    response.setCharacterEncoding("UTF-8");
    response.setContentType("text/html; charset=utf-8");
    PrintWriter writer=response.getWriter();
    //更新记录
    Connection conn=null;
    Statement stat=null;
    try{
      conn=Conn.getConnection();
      String sql=String.format("update result set name='%s', age=%d, result='%s' where id=%d", name, age, result, id);
      stat=conn.createStatement();
      stat.execute(sql);
      writer.append("修改成功!");
    } catch(ClassNotFoundException e){
      writer.append("修改失败!");
    } catch(SQLException e){
      writer.append("修改失败!");
    } finally{
      if(stat!=null){
        try{
          stat.close();
        } catch(SQLException e){
          e.printStackTrace();
        }
      }
      if(conn!=null){
        try{
          conn.close();
        } catch(SQLException e){
          e.printStackTrace();
        }
      }
    }
  }
```

步骤 6：制作成绩管理页面

（1）引入 style.css 样式和 index.js 文件，在 index.jsp 文件的<body>标签中编写<h1>、<table>、<button>，代码如下：

```
<%@ page language="java" contentType="text/html; charset=UTF-8"
  pageEncoding="UTF-8"%>
<%@page import="java.sql.*"%>
<%@ page import="com.stu.servlet.Conn"%>
<%
String account=null;
if(session.getAttribute("user.account")!=null){
  account=(String) session.getAttribute("user.account");
}
%>
<!DOCTYPE html>
<html>
<head>
<meta charset="UTF-8">
<link rel="stylesheet" type="text/css" href="css/style.css"/>
<script type="text/javascript" src="js/index.js"></script>
<title>学生成绩管理系统</title>
</head>
<body>
  <h1>学生成绩管理系统</h1>
  <%
  if(account!=null){
  %>
  <table>
    <%
    Connection conn=Conn.getConnection();
    Statement stat=conn.createStatement();
    String sql="select * from result";
    ResultSet rs=stat.executeQuery(sql);
    //遍历记录集
    while(rs.next()){
    %>
    <tr>
      <td><%=rs.getInt("id")%></td>
      <td><%=rs.getString("name")%></td>
      <td><%=rs.getInt("age")%></td>
      <td><%=rs.getString("result")%></td>
      <td>
        <button onclick="toUpdate(this)">修改</button>
        <button onclick="remove(this)">删除</button>
      </td>
    </tr>
```

```
    <%
    }
    rs.close();
    stat.close();
    conn.close();
    %>
    <tr>
      <td colspan="5"><a href="insert.jsp"><button>添加</button></a></td>
    </tr>
</table>
<%
} else{
%>
<button onclick="window.location.href='login.jsp'">登录</button>
<%
}
%>

<script type="text/javascript">
  function remove(ele){
    let id=ele.parentElement.parentElement.children[0].innerText;
    window.location.href="Remove?id="+id;
  }

  function toUpdate(ele){
    let id=ele.parentElement.parentElement.children[0].innerText;
    window.location.href="update.jsp?id="+id;
  }
</script>

</body>
</html>
```

运行结果如图 10-2 所示。

（2）引入 style.css 样式和 index.js 文件，在 insert.jsp 文件的<body>标签中编写<h1>、<table>，代码如下：

```
<%@ page language="java" contentType="text/html; charset=UTF-8"
    pageEncoding="UTF-8"%>
<!DOCTYPE html>
<html>
<head>
<meta charset="UTF-8">
    <link rel="stylesheet" type="text/css" href="css/style.css"/>
    <title>学生成绩管理系统</title>
</head>

<body>
```

```html
        <h1>学生成绩管理系统</h1>
        <table>
            <tr>
                <td>姓名:</td>
                <td><input type="text" name="name"></td>
            </tr>
            <tr>
                <td>年龄:</td>
                <td><input type="text" name="age"></td>
            </tr>
            <tr>
                <td>成绩:</td>
                <td><input type="text" name="result"></td>
            </tr>
            <tr>
                <td colspan="2"><button onclick="insert()">添加</button></td>
            </tr>
        </table>
        </form>
    </body>

    <script type="text/javascript">
    function insert(){
        let name=document.getElementsByName("name")[0].value;
        let age=document.getElementsByName("age")[0].value;
        let result=document.getElementsByName("result")[0].value;
        let ajax;
        if(window.XMLHttpRequest){
            ajax=new XMLHttpRequest();
        } else{
            ajax=new ActiveXObject("Microsoft.XMLHTTP");
        }
        ajax.open("POST", "Insert", false)
        ajax.setRequestHeader("Content-type", "application/x-www-form-urlencoded; charset=UTF-8");
        ajax.send("name="+name+"&age="+age+"&result="+result);
        alert(ajax.responseText);
        window.location.href="index.jsp";
    }
    </script>

</html>
```

运行结果如图 10-8 所示。

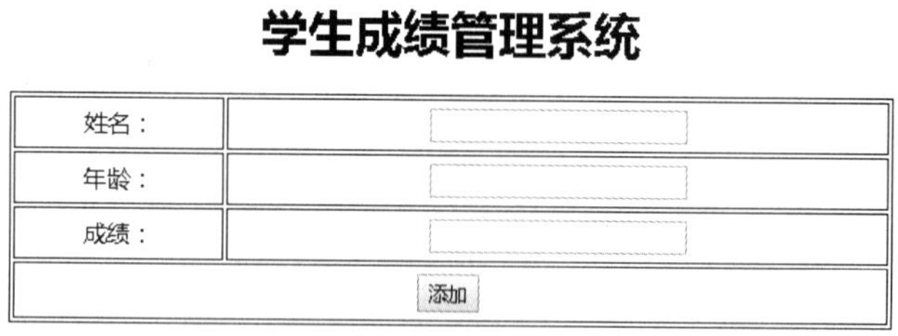

学生成绩管理系统

姓名：	
年龄：	
成绩：	
添加	

图 10-8

（3）引入 style.css 样式和 update.js 文件，在 update.Java 文件的<body>标签中编写<h1>、<table>，代码如下：

```
<%@ page language="java" contentType="text/html; charset=UTF-8"
    pageEncoding="UTF-8"%>
<!DOCTYPE html>
<html>
<head>
<meta charset="UTF-8">
<link rel="stylesheet" type="text/css" href="css/style.css"/>
<title>学生成绩管理系统</title>
</head>

<body>
  <h1>学生成绩管理系统</h1>
  <table>
    <tr>
      <td>姓名：</td>
      <td><input type="text" name="name"></td>
    </tr>
    <tr>
      <td>年龄：</td>
      <td><input type="text" name="age"></td>
    </tr>
    <tr>
      <td>成绩：</td>
      <td><input type="text" name="result"></td>
    </tr>
    <tr>
      <td colspan="2"><button onclick="update()">修改</button></td>
    </tr>
  </table>
</body>

<script type="text/javascript">
  let id;
  window.onload=function(){
    id=window.location.search.substring(1).split("=")[1];
    let ajax;
```

```
    if(window.XMLHttpRequest){
      ajax=new XMLHttpRequest();
    } else{
      ajax=new ActiveXObject("Microsoft.XMLHTTP");
    }
    ajax.open("GET", "Select?id="+id, false);
    ajax.send();
    let data=JSON.parse(ajax.responseText);
    document.getElementsByName("name")[0].value=data.name;
    document.getElementsByName("age")[0].value=data.age;
    document.getElementsByName("result")[0].value=data.result;
  }

  function update(){
    let name=document.getElementsByName("name")[0].value;
    let age=document.getElementsByName("age")[0].value;
    let result=document.getElementsByName("result")[0].value;
    let ajax;
    if(window.XMLHttpRequest){
      ajax=new XMLHttpRequest();
    } else{
      ajax=new ActiveXObject("Microsoft.XMLHTTP");
    }
    ajax.open("POST", "Update", false)
    ajax.setRequestHeader("Content-type",
        "application/x-www-form-urlencoded;charset=UTF-8");
    ajax.send("id="+id+"&name="+name+"&age="+age+"&result="+result);
    alert(ajax.responseText);
    window.location.href="index.jsp";
  }
</script>

</html>
```

运行结果如图 10-3 所示。

第 11 章 SSM 框架：第一个 SSM 程序

11.1 实验目标

（1）掌握 SSM 框架工程的创建方法，以及配置文件的编写方法。
（2）掌握 MyBatis 数据库信息的配置方法。
（3）掌握 MyBatis 数据库映射器的使用方法。
（4）掌握 Controller 控制器类的编写方法。
（5）掌握 @Autowire 注解。
（6）掌握 JSP 页面的编写方法。
（7）综合运用 SSM 框架，编写第一个 SSM 程序。

11.2 实验任务

搭建动态网站工程，使用 SSM 框架，完成 SSM 框架的集成和配置。该项目需要用到 MySQL 数据库，创建 first.sql 数据库脚本文件，使用 source 命令执行 first.sql 脚本文件创建数据库。first.sql 脚本内容如下：

```sql
DROP DATABASE IF EXISTS 'first';
CREATE DATABASE 'first';
USE 'first';

SET FOREIGN_KEY_CHECKS=0;

-- ----------------------------
-- Table structure for user
-- ----------------------------
DROP TABLE IF EXISTS 'user';
CREATE TABLE 'user'(
  'id' int(11) NOT NULL AUTO_INCREMENT,
  'name' varchar(255) NOT NULL,
```

```
    PRIMARY KEY ('id')
) ENGINE=InnoDB AUTO_INCREMENT=2 DEFAULT CHARSET=UTF8;

-- ----------------------------
-- Records of user
-- ----------------------------
INSERT INTO 'user' VALUES ('1', 'SSM');
```

数据库名为 first，包含 user 表，表中含有 id 和 name 两个字段，如图 11-1 所示。

图 11-1

使用 Tomcat 运行工程，显示从数据库 user 表中查询出来的用户名，如图 11-2 所示。

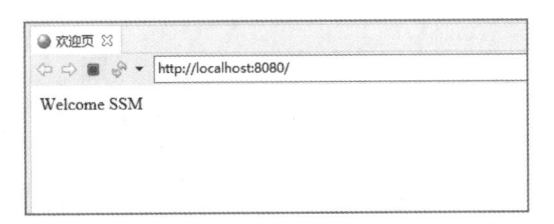

图 11-2

11.3 设计思路

1．工程设计

创建 Dynamic Web Project 项目——SSMFirst，集成 SSM 框架。

2．文件设计

文件设计如表 11-1 所示。

表 11-1

类型	文件/类	说明
Java 文件	com.first.UserMapper	User 数据库映射接口类
	com.first.WelcomeController	控制类文件
	com.first.User	User 实体类
JSP 文件	WEB-INF/jsp/welcome.jsp	欢迎页面
配置文件	src/db.properties	数据库配置文件
	src/applicationContext.xml	Spring 框架配置文件
	src/springmvc.xml	SpringMVC 框架配置文件
	WEB-INF/web.xml	网站主配置文件

3．页面设计

欢迎页面为 welcome.jsp，如图 11-3 所示。

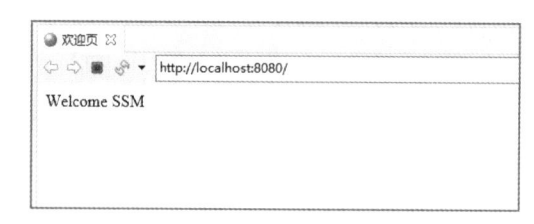

图 11-3

4．路由设计

（1）路由："/"。
（2）响应函数：WelcomeController 类、welcome 函数。

5．控制类

欢迎控制类：WelcomeController，使用@Controller 注解；包括 UserMapper 接口类对象，可进行数据库查询。

6．模型类

（1）用户实体类：User。
（2）用户数据库映射接口类：UserMapper，使用@Select 注解。

11.4 实验实施（跟我做）

步骤 1：创建工程

启动 Eclipse，选择"File"→"New"→"Dynamic Web Project"选项，在打开的向导窗口中的"Project name"文本框中输入"SSMFirst"工程名，在"Target runtime"下拉列表中选择"Apache Tomcat v8.5"选项，如图 11-4 所示。

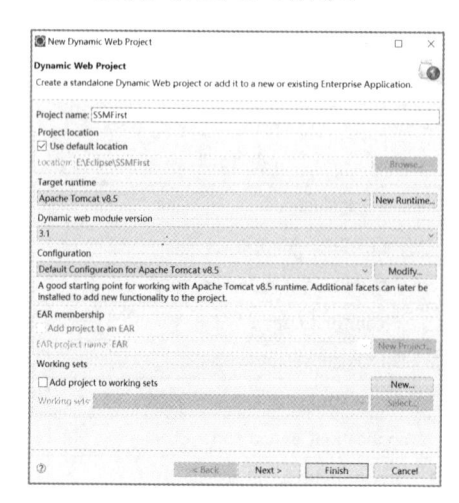

图 11-4

单击"Next"按钮，在打开的"Web Module"配置窗口中选中"Generate web.xml deployment descriptor"复选框，如图 11-5 所示，向导将自动创建 web.xml 文件。

图 11-5

单击"Finish"按钮，完成创建工程的操作。在 Package Explorer 中可以看到新创建的 SSMFirst 工程，如图 11-6 所示。

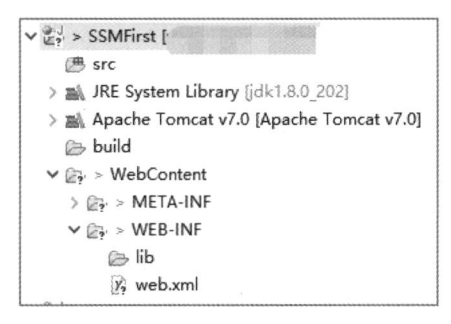

图 11-6

步骤 2：导入 SSM 包

1．下载 MyBatis 包

MyBatis 代码托管在 Github 上，MyBatis 的官网下载地址为 https://github.com/mybatis/mybatis-3/releases/tag/mybatis-3.5.6，如图 11-7 所示。

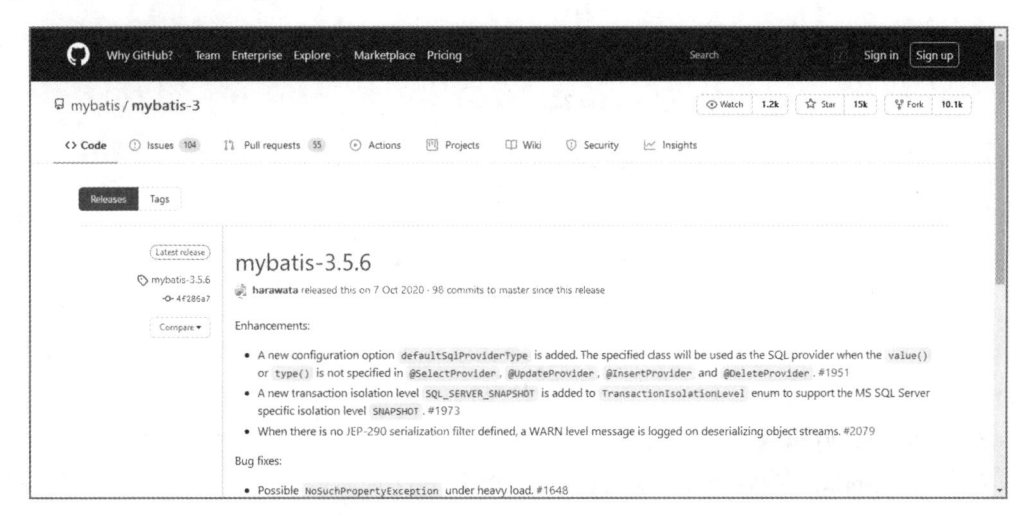

图 11-7

在页面下方找到 mybatis-3.5.6.zip 下载链接，如图 11-8 所示，单击链接进行下载。

图 11-8

下载成功后进行解压，mybatis-3.5.6.jar 包在根目录下，如图 11-9 所示，将该包复制到工程的"WebContent/WEB-INF/lib"文件夹中。

图 11-9

lib 文件夹中是相关的依赖包，将该文件夹中除"asm-7.1.jar"、"cglib-3.3.0.jar"和"ognl-3.2.15.jar"之外的所有 jar 包（图 11-10）都复制到工程的"WebContent/WEB-INF/lib"文件夹中。

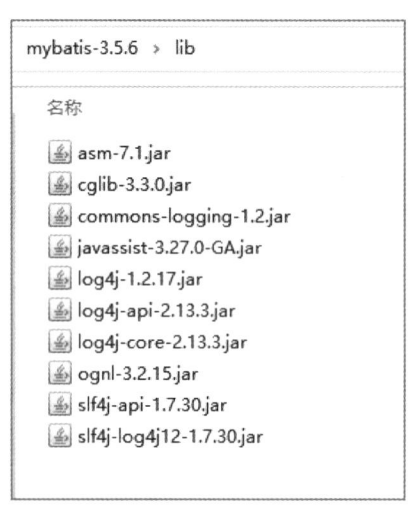

图 11-10

2. 下载 Spring 框架

首先进入 Spring 官网：https://spring.io/，如图 11-11 所示。

图 11-11

单击"Projects"下拉按钮,在弹出的下拉列表中选择"Spring Framework"选项,如图 11-12 所示,进入 Spring 框架主页面,如图 11-13 所示。

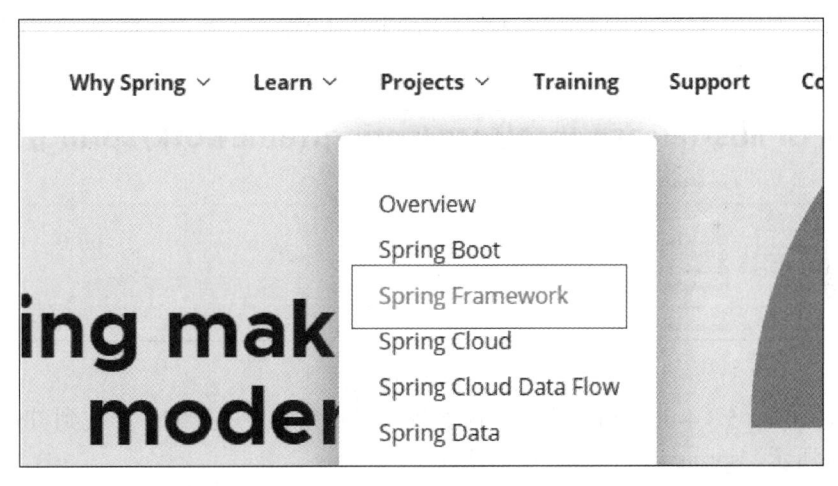

图 11-12

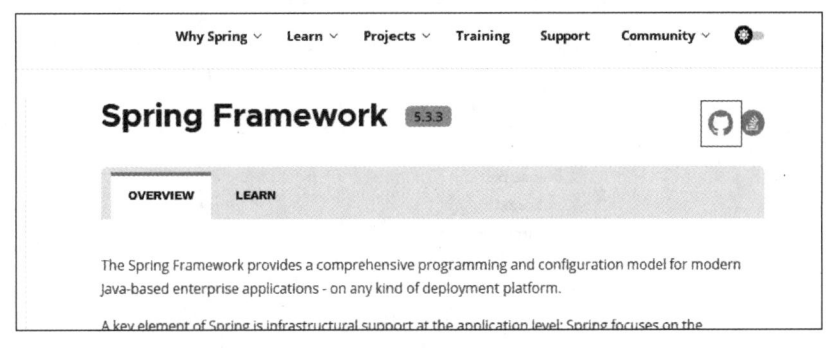

图 11-13

因为是托管在 Github 上的,所以单击 Github 标志,进入如图 11-14 所示的页面。

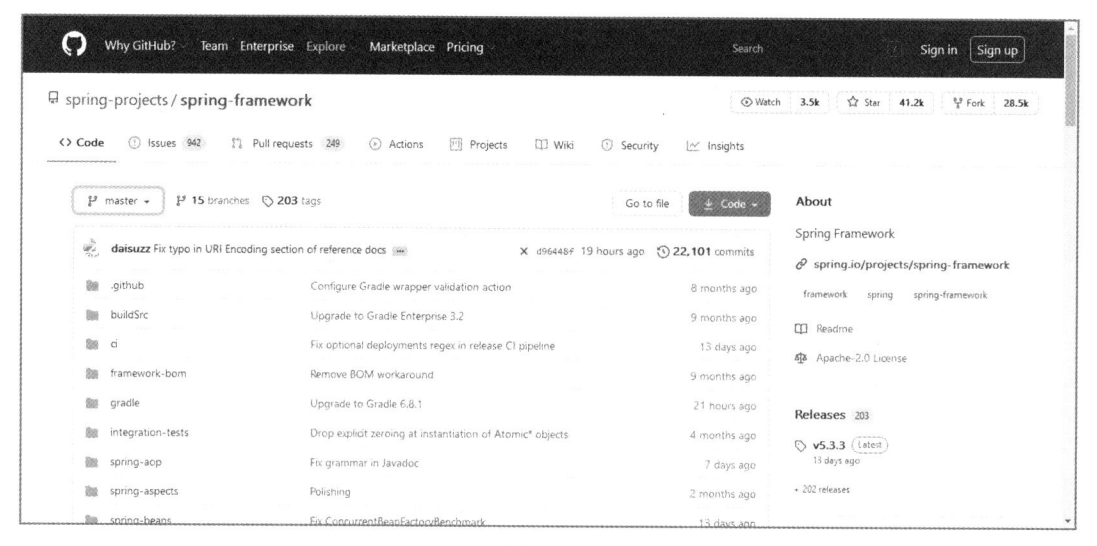

图 11-14

spring 框架的下载网址为 https://repo.spring.io/list/libs-release-local/org/springframework/spring/5.3.4/，如图 11-15 所示。

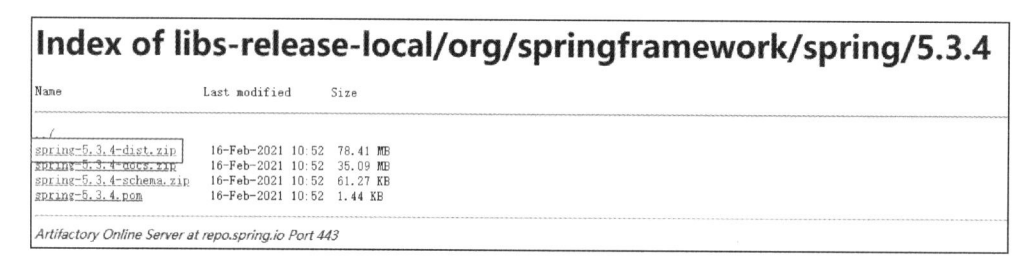

图 11-15

单击"spring-5.3.4-dist.zip"链接进行下载。下载完成后进行解压，找到 libs 文件夹，如图 11-16 所示，spring 的相关 jar 包都在其中。

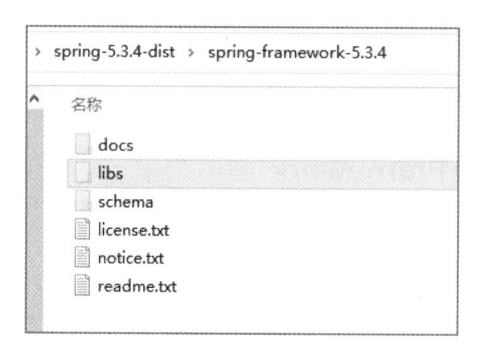

图 11-16

将如下 jar 包复制到工程的"WebContent/WEB-INF/lib"文件夹中，如图 11-17 所示。

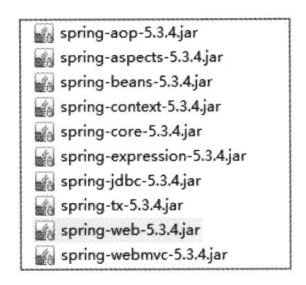

图 11-17

3．下载 Spring 和 MyBatis 集成包

MyBatis 和 Spring 框架集成的官网为 http://mybatis.org/spring/zh/index.html，如图 11-18 所示。

图 11-18

MyBatis 和 Spring 整合包的下载地址为 https://github.com/mybatis/spring/releases，如图 11-19 所示。

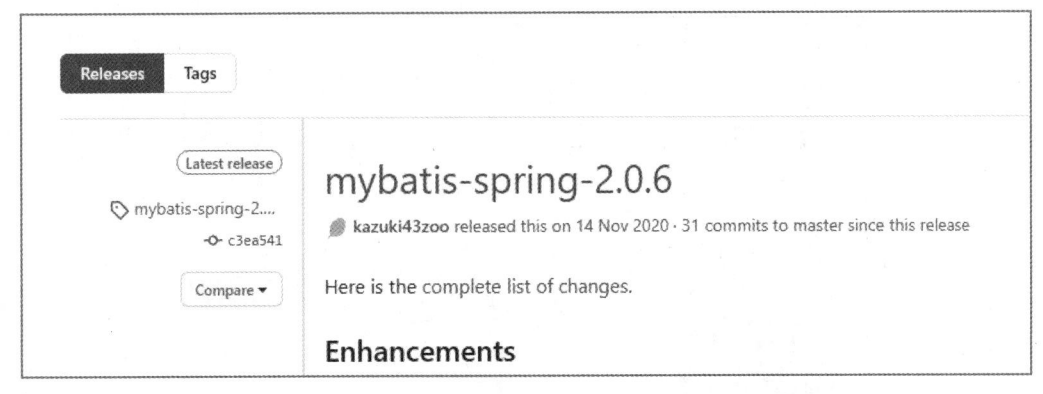

图 11-19

Github 上只提供了源码的下载，要下载 jar 包，可到 maven 上去搜索，地址为 https://search.maven.org/，进入网站后输入 mybatis-spring 进行搜索，如图 11-20 所示。

图 11-20

下载 mybatis-spring-2.0.6.jar 包，将文件复制到工程的"WebContent/WEB-INF/lib"文件夹中。

Spring 与 MyBatis 集成时除 Spring 和 MyBatis 相关包外，还需要一个 MySQL 数据库驱动包，其官网地址为 https://dev.mysql.com/downloads/connector/j/，如图 11-21 所示。

图 11-21

下载相应的 zip 文件，解压后将"mysql-connector-java-8.0.23.jar"文件复制到工程的"WebContent/WEB-INF/lib"文件夹中。

4．导入 jar 包

打开"WebContent/WEB-INF/lib"文件夹，选中所有的 jar 包，右击，在弹出的快捷菜单中选择"Build Path"→"Add to Build Path"选项，如图 11-22 所示，将 jar 包添加到工程中。

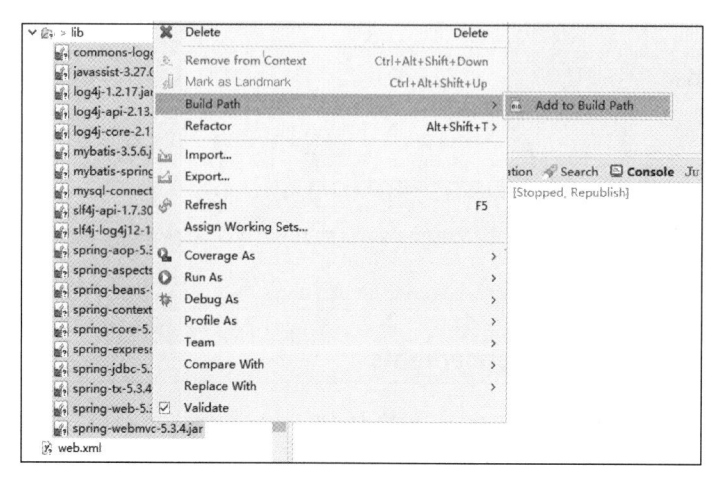

图 11-22

在工程的"Referenced Libraries"中可以看到添加的 jar 包，如图 11-23 所示。

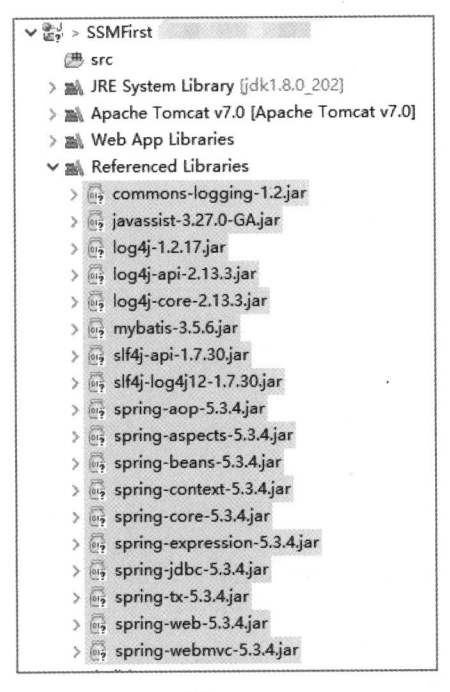

图 11-23

步骤 3：集成 SpringMVC

1. 配置 springmvc.xml

在 src 文件夹下创建 springmvc.xml 配置文件，用于配置 SpringMVC 框架。

```xml
<?xml version="1.0" encoding="UTF-8"?>
<beans xmlns="http://www.springframework.org/schema/beans"
    xmlns:xsi="http://www.w3.org/2001/XMLSchema-instance"
    xmlns:tx="http://www.springframework.org/schema/tx"
    xmlns:context="http://www.springframework.org/schema/context"
    xmlns:mvc="http://www.springframework.org/schema/mvc"
    xsi:schemaLocation="http://www.springframework.org/schema/beans
    http://www.springframework.org/schema/beans/spring-beans-3.0.xsd
    http://www.springframework.org/schema/tx
    http://www.springframework.org/schema/tx/spring-tx-3.0.xsd
    http://www.springframework.org/schema/context
    http://www.springframework.org/schema/context/spring-context-3.0.xsd
    http://www.springframework.org/schema/mvc
    http://www.springframework.org/schema/mvc/spring-mvc-3.0.xsd">

</beans>
```

（1）使用注解（annotation-driven）模式，自动扫描"com.first"包中的类。

```xml
<?xml version="1.0" encoding="UTF-8"?>
<beans ……>
```

```xml
<!-- 使用组件扫描 -->
<context:component-scan base-package="com.first"/>
</beans>
```

（2）将"/WEB-INF/jsp/"文件夹中的"jsp"文件作为视图，并在"WEB-INF"文件夹下创建"jsp"文件夹。

```xml
<?xml version="1.0" encoding="UTF-8"?>
<beans ……>
    <!-- 配置SpringMVC视图解析器 -->
    <bean
        class="org.springframework.web.servlet.view.InternalResourceViewResolver">
        <!-- 路径前缀 -->
        <property name="prefix" value="/WEB-INF/jsp/"/>
        <!-- 路径后缀 -->
        <property name="suffix" value=".jsp"/>
        <!-- 前缀+视图逻辑名+后缀=真实路径 -->
    </bean>
</beans>
```

（3）使用<mvc:annotation-driven>替代注解处理器和适配器的配置。

```xml
<?xml version="1.0" encoding="UTF-8"?>
<beans ……>
    <mvc:annotation-driven />
    <!-- 处理器映射器 -->
    <bean
        class="org.springframework.web.servlet.handler.BeanNameUrlHandlerMapping"/>
    <!-- 处理器适配器 -->
    <bean
        class="org.springframework.web.servlet.mvc.SimpleControllerHandlerAdapter"/>
</beans>
```

2. 修改 web.xml 配置文件

去掉 welcome 页面的相关代码。使用 DispatcherServlet 作为前端控制器，读取 springmvc.xml 配置文件，接管所有请求。

```xml
<?xml version="1.0" encoding="UTF-8"?>
<web-app xmlns:xsi="http://www.w3.org/2001/XMLSchema-instance" xmlns="http://java.sun.com/xml/ns/javaee" xsi:schemaLocation="http://java.sun.com/xml/ns/javaee http://java.sun.com/xml/ns/javaee/web-app_3_0.xsd" id="WebApp_ID" version="3.0">
    <display-name>SSMFirst</display-name>

    <!-- 配置DispatchcerServlet -->
    <servlet>
        <servlet-name>dispatcher</servlet-name>
        <servlet-class>org.springframework.web.servlet.DispatcherServlet</servlet-class>
        <init-param>
```

```xml
        <param-name>contextConfigLocation</param-name>
        <param-value>classpath:springmvc.xml</param-value>
    </init-param>
    <load-on-startup>1</load-on-startup>
  </servlet>
  <servlet-mapping>
    <servlet-name>dispatcher</servlet-name>
    <url-pattern>/</url-pattern>
  </servlet-mapping>

</web-app>
```

步骤 4：集成 MyBatis 框架

在"src"目录下创建配置文件 applicationContext.xml，将 MyBatis 与 Spring 进行整合，通过 Spring 管理 SqlSessionFactory、mapper 接口。

```xml
<?xml version="1.0" encoding="UTF-8"?>
<beans xmlns="http://www.springframework.org/schema/beans"
    xmlns:context="http://www.springframework.org/schema/context"
    xmlns:xsi="http://www.w3.org/2001/XMLSchema-instance"
    xmlns:aop="http://www.springframework.org/schema/aop"
    xmlns:tx="http://www.springframework.org/schema/tx"
    xmlns:p="http://www.springframework.org/schema/p"
    xsi:schemaLocation="
      http://www.springframework.org/schema/context
      http://www.springframework.org/schema/context/spring-context.xsd
      http://www.springframework.org/schema/beans
      http://www.springframework.org/schema/beans/spring-beans.xsd
      http://www.springframework.org/schema/tx
      http://www.springframework.org/schema/tx/spring-tx.xsd
      http://www.springframework.org/schema/aop
      http://www.springframework.org/schema/aop/spring-aop.xsd">

    //后面的配置添加在此处

</beans>
```

（1）添加数据库连接配置文件。在"src"文件夹下创建 db.properties 文件。

```
jdbc.driver=com.mysql.cj.jdbc.Driver
jdbc.url=jdbc:mysql://localhost:3306/first?useUnicode=true&characterEncoding=utf8
jdbc.username=root
jdbc.password=123456
```

（2）配置数据源（使用 Spring 提供的数据源）。

```xml
<?xml version="1.0" encoding="UTF-8"?>
<beans ……>
    <!-- 加载配置文件 -->
```

```xml
<context:property-placeholder location="classpath:db.properties"/>

    <!-- spring 数据源 -->
    <bean id="dataSource"
        class="org.springframework.jdbc.datasource.DriverManagerDataSource">
        <property name="driverClassName" value="${jdbc.driver}"/>
        <property name="url" value="${jdbc.url}"/>
        <property name="username" value="${jdbc.username}"/>
        <property name="password" value="${jdbc.password}"/>
    </bean>
</beans>
```

（3）配置 SqlSessionFactory。
```xml
<?xml version="1.0" encoding="UTF-8"?>
<beans ……>
    <!-- sqlSessinFactory -->
    <bean id="sqlSessionFactory"
        class="org.mybatis.spring.SqlSessionFactoryBean">
        <!-- 数据源,数据库连接池 -->
        <property name="dataSource" ref="dataSource"/>
        <!-- 给映射文件中的 resultType 设置别名 -->
        <property name="typeAliasesPackage" value="com.first"/>
    </bean>
</beans>
```

（4）配置 mapper 扫描器。
```xml
<?xml version="1.0" encoding="UTF-8"?>
<beans ……>
    <!-- mapper 扫描器 -->
    <!-- basePackage: 扫描包路径,中间可以用逗号或分号分隔定义多个包 -->
    <bean class="org.mybatis.spring.mapper.MapperScannerConfigurer">
        <property name="basePackage" value="com.first"></property>
        <property name="sqlSessionFactoryBeanName"
            value="sqlSessionFactory"/>
    </bean>
</beans>
```

（5）事务管理。
```xml
<?xml version="1.0" encoding="UTF-8"?>
<beans ……>
    <!-- 配置事务管理器 -->
    <bean id="transactionManager"
        class="org.springframework.jdbc.datasource.DataSourceTransactionManager">
        <property name="dataSource" ref="dataSource"/>
    </bean>
</beans>
```

步骤 5：加载 Spring 容器

在 web.xml 中，添加 Spring 容器监听器，加载 Spring 容器。加载 applicationContext.xml 配置文件。

```xml
<?xml version="1.0" encoding="UTF-8"?>
<web-app ……>
    <listener>
        <listener-class>org.springframework.web.context.ContextLoaderListener</listener-class>
    </listener>
    <!-- 加载spring容器 -->
    <context-param>
        <param-name>contextConfigLocation</param-name>
        <param-value>classpath:applicationContext.xml</param-value>
    </context-param>
    <listener>
        <listener-class>org.springframework.web.context.ContextLoaderListener</listener-class>
    </listener>
</web-app>
```

步骤 6：使用 MyBatis 操作数据库

1. 创建数据库

创建 first.sql 数据库脚本文件，编写如下创建数据库的脚本。

```sql
DROP DATABASE IF EXISTS `first`;
CREATE DATABASE 'first';
USE 'first';

SET FOREIGN_KEY_CHECKS=0;

-- ----------------------------
-- Table structure for user
-- ----------------------------
DROP TABLE IF EXISTS 'user';
CREATE TABLE 'user'(
  'id' int(11) NOT NULL AUTO_INCREMENT,
  'name' varchar(255) NOT NULL,
  PRIMARY KEY ('id')
) ENGINE=InnoDB AUTO_INCREMENT=2 DEFAULT CHARSET=UTF8;

-- ----------------------------
-- Records of user
-- ----------------------------
INSERT INTO 'user' VALUES ('1', 'SSM');
```

使用 source 命名运行脚本，创建 first 数据库，其中创建了一个 user 表，里面有一个 id 为自增主键，name 为名称。并在创建完数据库后向 user 表中插入一条记录。

2. 创建实体类 User

在"com.first"包中创建 User 类，添加 id 和 name 属性，并添加 setter 与 getter 方法，添加@Alias 注解，给类起一个别名"User"。

```java
package com.first;

import org.apache.ibatis.type.Alias;

@Alias("User")
public class User{
  private int id;
  private String name;

  public int getId(){
    return id;
  }

  public void setId(int id){
    this.id=id;
  }

  public String getName(){
    return name;
  }

  public void setName(String name){
    this.name=name;
  }
}
```

3. 创建 UserMapper 接口类

在"com.first"包中创建 UserMapper 接口类，添加 getUserName( )函数。

```java
package com.first;

import org.apache.ibatis.annotations.Select;

public interface UserMapper{
  @Select("select name from user where id=#{id}")
  public String getUserName(int id);
}
```

步骤 7：编写 WelcomeController 类

1. 添加 welcome.jsp 页面

在"WEB-INF/jsp"文件夹下创建 welcome.jsp 文件，将编码格式修改为 UTF-8，将标题改为"欢迎页"。

```jsp
<%@ page language="java" contentType="text/html; charset=UTF-8"
    pageEncoding="UTF-8"%>
<!DOCTYPE html>
<html>
<head>
<meta charset="UTF-8">
<title>欢迎页</title>
</head>
<body>
Welcome
</body>
</html>
```

2. 创建 WelcomeController 类

在 "com.first" 包下创建 WelcomeController 类，使用@Controller 指定该类为控制器类。定义方法为 welcome 函数，使用@RequestMapping 注解配置 GET 路由 "/" 访问该函数。

```java
package com.first;

import org.springframework.stereotype.Controller;
import org.springframework.web.bind.annotation.RequestMapping;
import org.springframework.web.bind.annotation.RequestMethod;

@Controller
public class WelcomeController{

  @RequestMapping(value="/", method=RequestMethod.GET)
  public String welcome(){
    return "welcome";
  }
}
```

3. 添加 UserMapper 属性

在 WelcomeController 类中添加 UserMapper 对象，使用@Autowired 注解自动加载对象。

```java
@Controller
public class WelcomeController{
  @Autowired
  private UserMapper userMapper;

}
```

4. 编写 welcome 函数

查询 id 为 1 的用户名，并返回 "welcome" 页面。

```java
@Controller
public class WelcomeController{
  ……
  @RequestMapping(value="/", method=RequestMethod.GET)
```

```java
public String welcome(Model model){
  String name=this.userMapper.getUserName(1);
  model.addAttribute("name", name);
  return "welcome";
  }
}
```

5. 修改 welcome.jsp 页面

从 request 中取出属性，并在页面显示。

```jsp
<%
String name=(String)request.getAttribute("name");
%>
<!DOCTYPE html>
<html>
……
<body>
Welcome <%=name %>
</body>
</html>
```

步骤 8：发布和运行

1. 将项目发布到 Tomcat

打开 Servers 视窗，右击 Tomcat 服务器，在弹出的快捷菜单中选择"Add and Remove"选项，如图 11-24 所示。

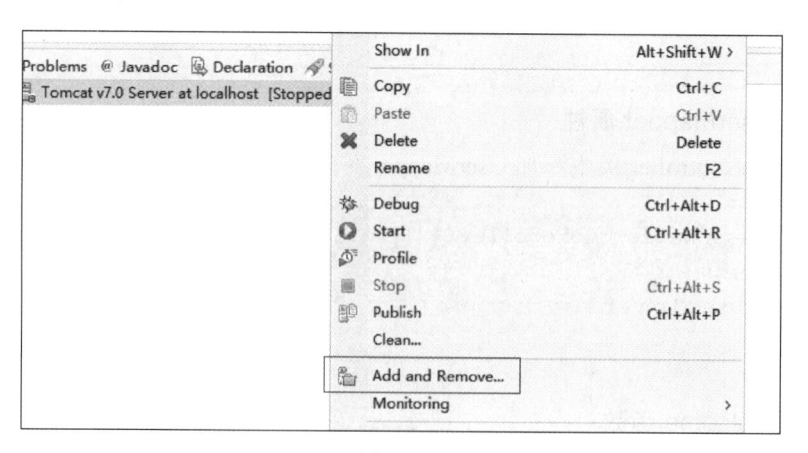

图 11-24

弹出选择项目对话框，如图 11-25 所示。

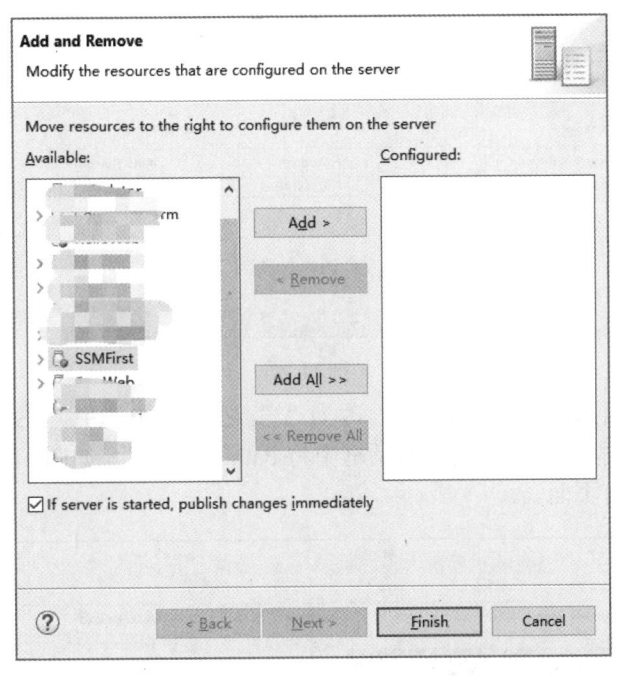

图 11-25

单击"Add"按钮将选择的项目添加到右侧的列表框中，如图 11-26 所示，然后单击"Finish"按钮即可。

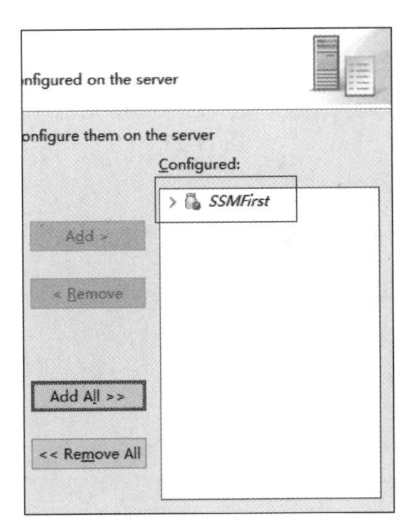

图 11-26

2．设置根目录

在 Servers 视窗中双击 Tomcat 服务器，将显示的窗口切换到"Web Modules"页面，如图 11-27 所示。

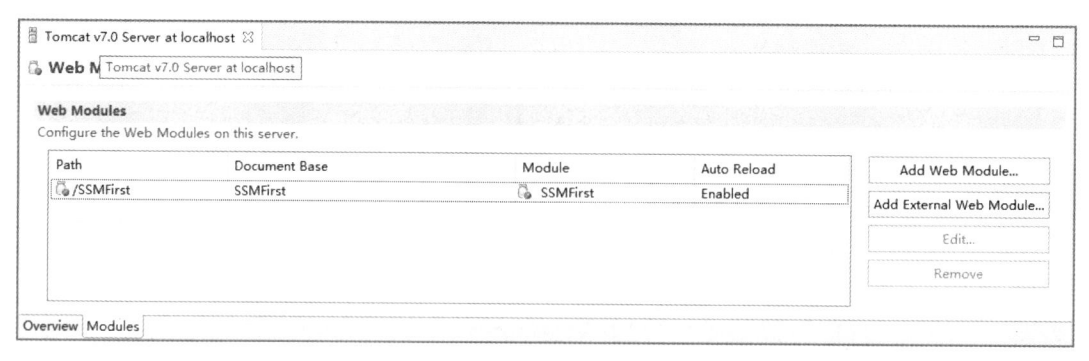

图 11-27

默认项目的根目录为"/SSMFirst",改为"/"。在表格中选中"SSMFirst"选项,然后单击右侧的"Edit"按钮。在弹出的对话框中将 Path 改为"/",如图 11-28 所示,然后单击"OK"按钮保存当前页面。

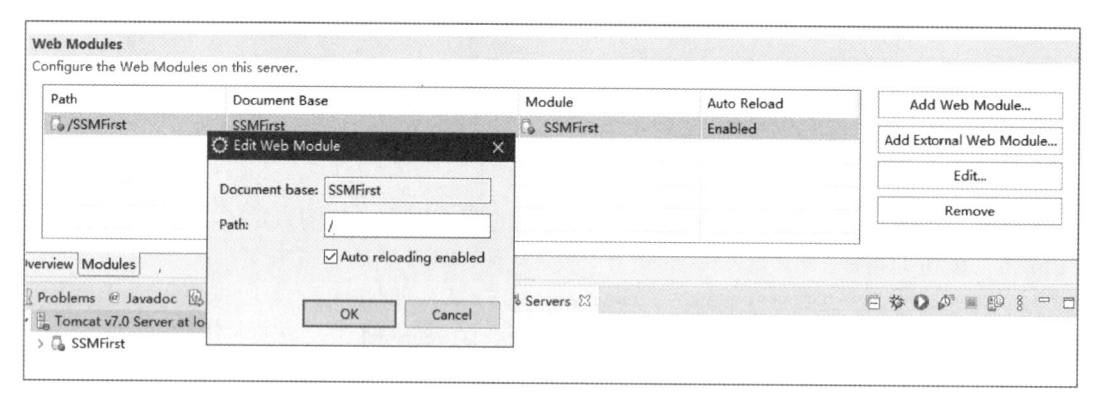

图 11-28

3. 运行

选中程序,右击,在弹出的快捷菜单中选择"Run As"→"Run on Server"选项,即可运行程序,如图 11-3 所示。

第 12 章
SSM 框架：在线答题

12.1 实验目标

（1）掌握 Spring 与 SpringMVC 框架的集成方法。
（2）掌握 Controller 控制器类的编写方法。
（3）掌握 ArrayList 的使用方法。
（4）掌握 static 成员变量。
（5）掌握 JSP 页面的编写方法。
（6）综合运用 SSM 框架，编写"在线答题"程序。

12.2 实验任务

使用 Spring 与 SpringMVC 框架编写一个简单的在线答题程序。

（1）试题共 4 道数学题，每题 3 个选项，都为单选题。每题 25 分，共 100 分。题目内容、选项和答案如下。

第 1 题，10+4=?；选项，A.12　B.14　C.16；答案，B。
第 2 题，20-9=?；选项，A.7　B.13　C.11；答案，C。
第 3 题，7×3=?；选项，A.21　B.24　C.25；答案，A。
第 4 题，8÷2=?；选项，A.10　B.2　C.4；答案，C。

（2）每做完一题，单击"下一题"按钮，提交当前题目答案，并显示下一题内容，如图 12-1 所示。

图 12-1

(3)在最后一题时,按钮显示为"提交"。
(4)单击"提交"按钮后,显示答对的题目数量和得分情况,总分数为 100 分,如图 12-2 所示。

图 12-2

12.3 设计思路

1. 工程设计

创建 Dynamic Web Project 项目——Quiz,使用 Spring+SpringMVC 框架。

2. 文件设计

文件设计如表 12-1 所示。

表 12-1

类型	文件/类	说明
Java 文件	com.qz.QuizController	控制类文件
	com.qz.Question	题目模型类
JSP 文件	WEB-INF/jsp/quiz.jsp	答题页
	WEB-INF/jsp/result.jsp	结果页
配置文件	src/springmvc.xml	SpringMVC 框架配置文件
	src/applicationContext.xml	Spring 框架配置文件
	WEB-INF/web.xml	网站主配置文件

3. 页面设计

(1)答题页面为 quiz.jsp。
(2)结果页面为 result.jsp,如图 12-3 所示。

4. 路由设计

(1)进入答题系统路由(GET),请求方式为 GET;URL 为/;响应函数为 QuizController::start( )。
(2)提交当前题答案,进入下一题(POST),请求方式为 POST;URL 为"/quiz/next/{qid}";响应函数为 QuizController::next( )。
(3)提交最后一道题,并显示答题结果(POST),请求方式为 POST;URL 为"/quiz/submit";响应函数为 QuizController::submit( )。

在线答题

题号qid —— 第2题
题干stem —— 20 - 9 = ?
选项options —— ○ A.7
　　　　　　　○ B.13
　　　　　　　○ C.11
　　　　　　　[下一题]

最后一题时，按钮显示提交

在线答题

答题结束
共答对4题，获得100分 —— 答题结果
[重做]

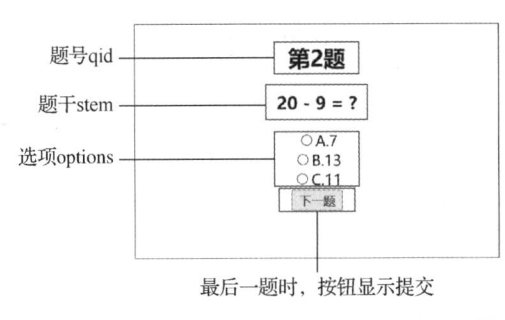

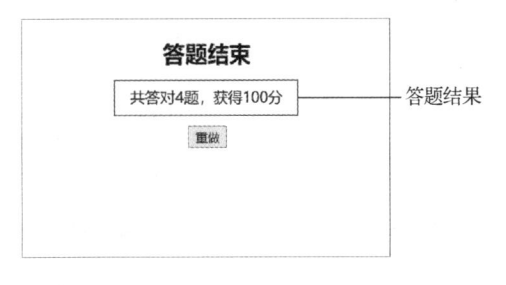

图 12-3

5．控制类

（1）答题控制类：QuizController。其继承于 Controller 类。

（2）使用@Controller 注解指定该类为控制器类，在 springmvc.xml 文件中设置 "<context:component-scan base-package="com.qz"/>"，程序启动时自动扫描 "com.qz" 下的类。

（3）开始答题函数。

`public String start(HttpSession session, Model model)`

（4）获得下一题，并保存当前题的答案。

`public String next(@PathVariable(name="qid") int qid, HttpSession session, HttpServletRequest request, Model model)`

（5）提交试卷，计算分数，并返回结果。

`public String submit(HttpSession session, HttpServletRequest request, Model model)`

（6）通过题号获得试题信息。

`private Question getQuestion(int qid)`

6．数据定义

（1）在 QuizController 类中定义 private static String[][] questions 数组对象。

（2）使用二维数组保存试题数据。

（3）定义 private static final String PARAM_ANSWERS="answers"，作为 Session 中保存用户答案的对象的键值。

12.4 实验实施（跟我做）

步骤1：创建工程

启动 Eclipse，选择 "File" → "New" → "Dynamic Web Project" 选项，在打开的向导窗口中的 "Project name" 文本框中输入 "Quiz" 工程名，在 "Target runtime" 下拉列表中选择 "Apache Tomcat v8.5" 选项。

单击 "Next" 按钮，在打开的 "Web Module" 配置窗口中选中 "Generate web.xml deployment descriptor" 复选框，向导将自动创建 web.xml 文件。

单击"Finish"按钮，完成创建工程的操作。在 Package Explorer 中可以看到新创建的 Quiz 工程。

导入 Spring 和 SpringMVC 框架的包，以及依赖 commons-logging-1.2.jar 包，如图 12-4 所示。

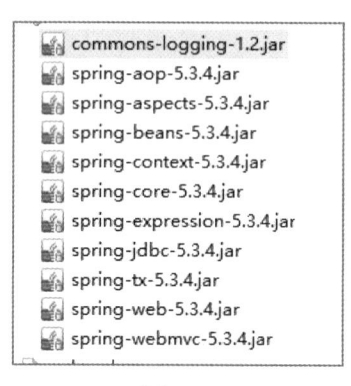

```
commons-logging-1.2.jar
spring-aop-5.3.4.jar
spring-aspects-5.3.4.jar
spring-beans-5.3.4.jar
spring-context-5.3.4.jar
spring-core-5.3.4.jar
spring-expression-5.3.4.jar
spring-jdbc-5.3.4.jar
spring-tx-5.3.4.jar
spring-web-5.3.4.jar
spring-webmvc-5.3.4.jar
```

图 12-4

在 src 下创建 applicationContext.xml 配置文件。

```xml
<?xml version="1.0" encoding="UTF-8"?>
<beans xmlns="http://www.springframework.org/schema/beans"
 xmlns:context="http://www.springframework.org/schema/context"
 xmlns:xsi="http://www.w3.org/2001/XMLSchema-instance"
 xmlns:aop="http://www.springframework.org/schema/aop"
 xmlns:tx="http://www.springframework.org/schema/tx"
 xmlns:p="http://www.springframework.org/schema/p"
 xsi:schemaLocation="
     http://www.springframework.org/schema/context
     http://www.springframework.org/schema/context/spring-context.xsd
     http://www.springframework.org/schema/beans
     http://www.springframework.org/schema/beans/spring-beans.xsd
     http://www.springframework.org/schema/tx
     http://www.springframework.org/schema/tx/spring-tx.xsd
     http://www.springframework.org/schema/aop
     http://www.springframework.org/schema/aop/spring-aop.xsd">

</beans>
```

在 src 下创建 springmvc.xml 配置文件。

```xml
<?xml version="1.0" encoding="UTF-8"?>
<beans xmlns="http://www.springframework.org/schema/beans"
 xmlns:xsi="http://www.w3.org/2001/XMLSchema-instance"
 xmlns:tx="http://www.springframework.org/schema/tx"
 xmlns:context="http://www.springframework.org/schema/context"
 xmlns:mvc="http://www.springframework.org/schema/mvc"
 xsi:schemaLocation="http://www.springframework.org/schema/beans
   http://www.springframework.org/schema/beans/spring-beans-3.0.xsd
   http://www.springframework.org/schema/tx
   http://www.springframework.org/schema/tx/spring-tx-3.0.xsd
```

```xml
      http://www.springframework.org/schema/context
      http://www.springframework.org/schema/context/spring-context-3.0.xsd
      http://www.springframework.org/schema/mvc
      http://www.springframework.org/schema/mvc/spring-mvc-3.0.xsd">

    <mvc:annotation-driven/>

    <!-- 处理器映射器 -->
    <bean
      class="org.springframework.web.servlet.handler.BeanNameUrlHandlerMapping"/>

    <!-- 处理器适配器 -->
    <bean
      class="org.springframework.web.servlet.mvc.SimpleControllerHandlerAdapter"/>

    <!-- 使用组件扫描 -->
    <context:component-scan base-package="com.qz"/>

    <!-- 配置SpringMVC视图解析器 -->
    <bean
      class="org.springframework.web.servlet.view.InternalResourceViewResolver">
      <!-- 路径前缀 -->
      <property name="prefix" value="/WEB-INF/jsp/"/>
      <!-- 路径后缀 -->
      <property name="suffix" value=".jsp"/>
      <!-- 前缀+视图逻辑名+后缀=真实路径 -->
    </bean>

</beans>
```

编写 web.xml 配置文件。

```xml
<?xml version="1.0" encoding="UTF-8"?>
<web-app xmlns:xsi="http://www.w3.org/2001/XMLSchema-instance"
  xmlns="http://java.sun.com/xml/ns/javaee"
  xsi:schemaLocation="http://java.sun.com/xml/ns/javaee http://java.sun.com/xml/ns/javaee/web-app_3_0.xsd"
  id="WebApp_ID" version="3.0">
  <display-name>Quiz</display-name>
  <listener>
    <listener-class>org.springframework.web.context.ContextLoaderListener</listener-class>
  </listener>

  <!-- 加载spring容器 -->
  <context-param>
    <param-name>contextConfigLocation</param-name>
    <param-value>classpath:applicationContext.xml</param-value>
  </context-param>
```

```xml
<listener>
    <listener-class>org.springframework.web.context.ContextLoaderListener</listener-class>
</listener>

<!-- 配置 DispatchcerServlet -->
<servlet>
  <servlet-name>dispatcher</servlet-name>
  <servlet-class>org.springframework.web.servlet.DispatcherServlet</servlet-class>
    <init-param>
      <param-name>contextConfigLocation</param-name>
      <param-value>classpath:springmvc.xml</param-value>
    </init-param>
    <load-on-startup>1</load-on-startup>
</servlet>
<servlet-mapping>
    <servlet-name>dispatcher</servlet-name>
    <url-pattern>/</url-pattern>
</servlet-mapping>

<!-- 指明对于如上资源文件不采用 spring 的过滤器 -->
<servlet-mapping>
  <servlet-name>default</servlet-name>
  <url-pattern>*.css</url-pattern>
</servlet-mapping>

</web-app>
```

步骤 2：创建控制器类和路由函数

创建 com.qz 包，在包中创建 QuizConroller 类。给 QuizController 添加@Controller 注解，添加如下函数，并设置路由。

（1）进入答题系统路由（GET）：路由 "/" 调用 QuizController::start( )函数。

（2）提交当前题答案，进入下一题（POST）：路由 "/quiz/next/{qid}" 调用 QuizController@next( )函数。

（3）提交最后一道题，并显示答题结果（POST）：路由 "/quiz/submit" 调用 QuizController@submit( )函数。

```java
package com.qz;

……

@Controller
public class QuizController{
  @GetMapping("/")
  public String start(HttpSession session, Model model){
```

```java
    }

    @PostMapping("/quiz/next/{qid}")
    public String next(@PathVariable(name="qid") int qid, HttpSession session,
HttpServletRequest request, Model model){

    }

    @PostMapping("/quiz/submit")
    public String submit(HttpSession session, HttpServletRequest request,
Model model){

    }
}
```

步骤 3：创建 quiz.jsp 文件

（1）创建页面样式文件。

在"WebContent"下创建 css 文件夹，在 css 文件夹中创建 quiz.css 文件。

```css
h1{text-align: center;}
.box{
    margin: auto;
    border: solid 1px black;
    margin-top: 5%;
    width: 400px;
    height: 250px;
    text-align: center;
}
```

（2）在"WEB-INF/jsp/"中创建 quiz.jsp 文件。

导入静态 css/quiz.css 文件，并将编码格式修改为"UTF-8"。

```jsp
<%@ page language="java" contentType="text/html; charset=UTF-8"
    pageEncoding="UTF-8"%>
<!DOCTYPE html>
<html>
<head>
<meta charset="utf-8"/>
<link rel="stylesheet" href="/css/quiz.css">
</head>
<body>
    <h1>在线答题</h1>
    <div class="box">
    </div>
    </div>
</body>
</html>
```

显示当前的题号，添加 form 表单，判断当前题是否为最后一题，如果不是最后一题则 action 为/quiz/next/题号，如果是最后一题则为/quiz/submit。显示题干$stem，使用 foreach 显示选项，使用 if-else 显示不同的按钮。

```jsp
<%@ page language="java" contentType="text/html; charset=UTF-8"
  pageEncoding="UTF-8"%>
<%@ page import="com.qz.Question" %>
<%
Question question=(Question)request.getAttribute("question");
%>
<!DOCTYPE html>
<html>
<head>
<meta charset="utf-8"/>
<link rel="stylesheet" href="/css/quiz.css">
</head>
<body>
  <h1>在线答题</h1>
  <div class="box">
    <div>
      <h2 id="test_status">第<%=question.getId()%>题</h2>
      <div id="test">
      <%if(question.isLast()){%>
        <form method="post" action="/quiz/submit">
          <h3><%=question.getStem()%></h3>
          <%
          String[] options=question.getOptions();
          for(int i=0; i<3; i++){
            String option=options[i];
            String value=String.format("%c", 'A'+i);
          %>
            <input type="radio" name="choices" value="<%=value %>"> <%= option %> </input><br/>
          <%}%>
          <button type="submit">提交</button>
        </form>
      <%}else{ %>
        <form method="post" action="/quiz/next/<%=question.getId()%>">
          <h3><%=question.getStem()%></h3>
          <%
          String[] options=question.getOptions();
          for(int i=0; i<3; i++){
            String option=options[i];
            String value=String.format("%c", 'A'+i);
          %>
            <input type="radio" name="choices" value="<%=value %>"> <%= option %></input> <br/>
```

```
            <%}%>
            <button type="submit">下一题</button>
          </form>
      <%}%>
        </div>
      </div>
  </div>
</body>
</html>
```

步骤 4：创建 result.jsp 文件

在"WEB-INF/jsp"文件夹中创建 result.jsp 文件。

（1）显示答对的题目数量和得分。

设置页面编码格式为"UTF-8"，从 request 中获得传过来的 right_num 和 score 值，并在页面中显示。

```
<%@ page language="java" contentType="text/html; charset=UTF-8"
 pageEncoding="UTF-8"%>
<%
Integer rightNum=(Integer) request.getAttribute("right_num");
Integer score=(Integer) request.getAttribute("score");
%>
<!DOCTYPE html>
<html>
<head>
<meta charset="utf-8"/>
<link rel="stylesheet" href="/css/quiz.css">
</head>
<body>
  <h1>在线答题</h1>
  <div class="box">
    <div>
      <h2 id="test_status">答题结束</h2>
      <div id="test">共答对<%=rightNum%>题，获得<%=score%>分</div>
    </div>
  </div>
</body>
</html>
```

（2）添加"重做"按钮，返回"/"页面。

```
<!DOCTYPE html>
<html>
    ……
    <body>
      ……
        <div>
          ……
```

```
            <br/>
            <button type="button" onclick="window.location='/';">重做</button>
        </div>
    </body>
</html>
```

步骤 5：创建 QuizController 处理函数

(1) 定义试题数据。

```
@Controller
public class QuizController{
  //类中的静态成员
  private static String[][] questions={{"10+4=?", "12", "14", "16", "B"},
      {"20-9=?", "7", "13", "11", "C"}, {"7×3=?", "21", "24", "25", "A"},
      {"8÷2=?", "10", "2", "4", "C"}};
}
```

(2) 定义保存到 Session 中的属性名常量。

```
@Controller
public class QuizController{
  //类的常量变量
  private static final String PARAM_ANSWERS="answers";
}
```

(3) 创建 getQuestion( )函数，访问权限为 private，参数为题号，读取试题内容。

使用 for 循环解析出各个选项，在选项前添加字母 A～C。选项字符串格式为[A～C].选项内容。创建 Question 对象，设置 qid 题号，stem 题干，options 选项列表，last 是否为最后一道题的标志（若为最后一道题则为 true，否则为 false）。

```
@Controller
public class QuizController{

  /**
   * 通过试题索引获得试题
   *
   * @param qid
   * @return
   */
  private Question getQuestion(int qid){
    //获得试题信息
    String[] question=questions[qid];

    //创建试题对象
    Question q=new Question();
    q.setId(qid+1);            //设置试题 id
    q.setStem(question[0]);    //设置题干
    //设置选项,在选项前添加 A、B、C
    String[] options=new String[3];
    for(int i=1; i<4; i++){
```

```
        options[i-1]=String.format("%c.%s", 'A'+i-1, question[i]);
    }
    q.setOptions(options);
    //设置当前试题是否为最后一题
    q.setLast(questions.length==qid+1);
    //返回试题对象
    return q;
  }
}
```

（4）编写 start( )函数。

读取题目内容后，通过 model 传给 quiz 页面。

在函数参数中传入 Session 对象：HttpSession session。

调用 Session 中的 setAttribute( )函数设置"answers"列表对象。定义一个空的 List<Sring> 对象来保存用户填写的答案。

```
@Controller
public class QuizController{

  @GetMapping("/")
  public String start(HttpSession session, Model model){
    //读取第一题
    Question question=this.getQuestion(0);
    //创建用来保存用户答案的属性
    List<String>answers=new ArrayList<String>();
    session.setAttribute(PARAM_ANSWERS, answers);
    //显示 quiz.jsp 页面
    model.addAttribute("question", question);
    return "quiz";
  }
}
```

（5）编写 next( )函数。

next( )函数中的第 1 个参数从 URL 路径中获得，使用@PathVariable 注解定义，第 2 个参数为 Session 对象，第 3 个参数为请求对象 request。

从请求中取出用户的选择值，从 Session 中取出用户的答案数组，将该答案添加进去，并更新 Session 中的值。

```
@Controller
public class QuizController{

  @PostMapping("/quiz/next/{qid}")
  public String next(@PathVariable(name="qid") int qid, HttpSession session,
HttpServletRequest request,
    Model model){
    //获得上一题用户的答案
    String choice=request.getParameter("choices");
    //将用户的答案保存到会话中
    List<String> answers=(List<String>) session.getAttribute(PARAM_ANSWERS);
```

```
        answers.add(choice);
        session.setAttribute(PARAM_ANSWERS, answers);
        //获得下一题的内容
        Question question=this.getQuestion(qid);
        model.addAttribute("question", question);
        return "quiz";
    }
}
```

（6）编写 submit( )函数。

从 Session 中取出用户的答案数组，然后从 request 中获得最后一道题的答案并追加到答案数组中。定义 Model model 对象，用来将数据传递到页面中。

使用 count( )函数计算数组的元素个数。对比两个数组答案的匹配情况，计算出答对的题目数量，计算得分（100 分制）。

返回 result.jsp 页面，将分数和答对的题目数量传入。

```
@Controller
public class QuizController{

    @PostMapping("/quiz/submit")
    public String submit(HttpSession session, HttpServletRequest request, Model model){
        //从会话中取出前面的答案,并清空
        List<String>answers=(List<String>) session.getAttribute(PARAM_ANSWERS);
        session.removeAttribute(PARAM_ANSWERS);
        //获得最后一题用户的答案,更新答案列表
        String choice=request.getParameter("choices");
        answers.add(choice);
        //计算答对的题目的数量
        int questionCount=questions.length;
        int rightNum=0;
        for(int i=0; i<questionCount; i++){
            if(answers.get(i).equals(questions[i][4])){
                rightNum++;
            }

        }
        //计算分数
        int score=(int)(100*((double)rightNum/questionCount));

        //返回 result.jsp 页面
        model.addAttribute("score", score);
        model.addAttribute("right_num", rightNum);
        return "result";
    }
}
```

第 13 章

SSM 框架：个人博客

13.1 实验目标

（1）掌握 SSM 框架工程的创建方法，以及配置文件的编写方法。
（2）掌握 MyBatis 数据库信息的配置方法。
（3）掌握 MyBatis 数据库映射器的使用方法。
（4）掌握 Controller 控制器类的编写方法。
（5）掌握@Autowire 注解。
（6）掌握 JSP 页面的编写方法。
（7）综合运用 SSM 框架，编写"个人博客"程序。

13.2 实验任务

使用 SSM 框架编写一个简易的个人博客程序。

1. 登录与退出

（1）用户输入账号和密码，单击"登录"按钮，验证不成功，则显示"用户名或密码不一致"，如图 13-1 所示；若验证成功，则跳到主页。

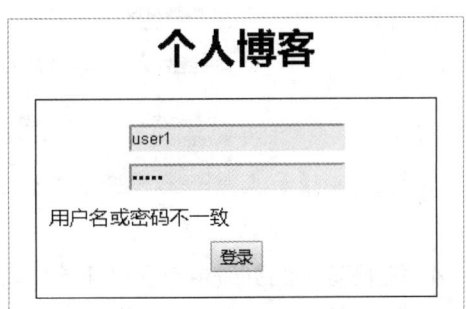

图 13-1

（2）主页显示登录用户名，如图 13-2 所示。单击账号超链接可以退出程序，然后主页上显示"未登录"，单击"未登录"超链接即可进入登录页面。

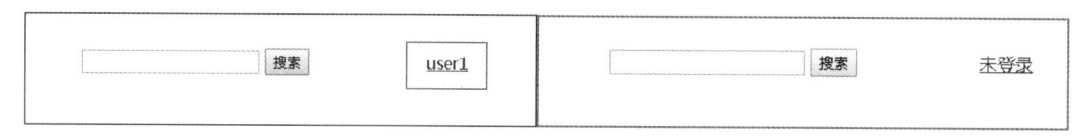

图 13-2

2. 添加博客

用户登录后,可以发布一条新的博客。输入标题和内容,单击"发布"按钮,即可添加博客,如图 13-3 所示。

图 13-3

3. 显示博客列表

(1)在左侧栏显示博客标题列表。
(2)在添加表单的下方显示博客详细列表,如图 13-4 所示。

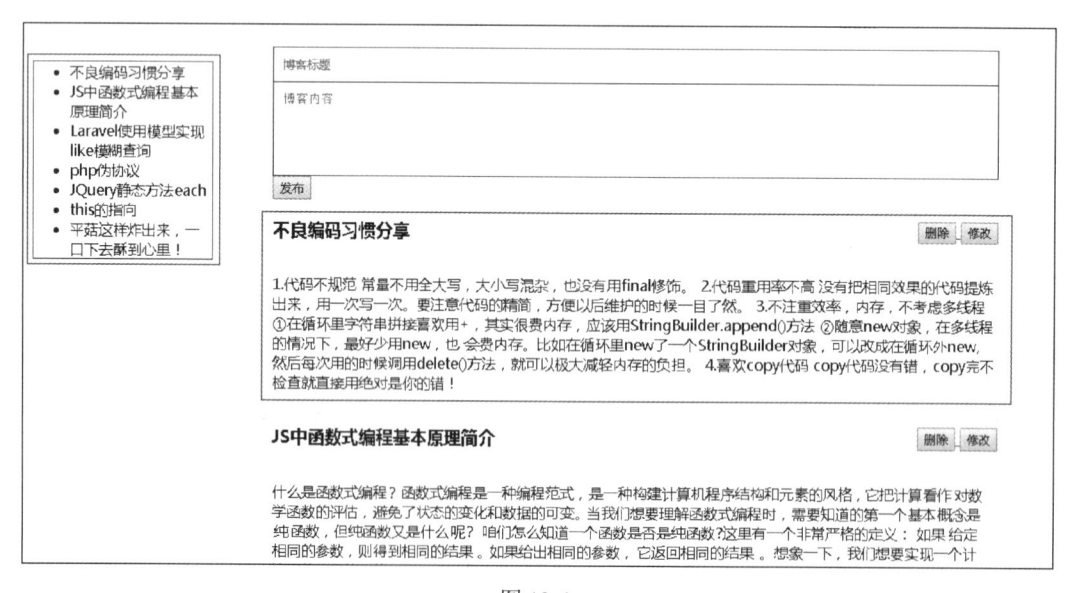

图 13-4

4. 在博客详细列表中包含"删除"与"修改"按钮

(1)用户登录后,单击"删除"按钮,即可删除此条记录。
(2)用户登录后,单击"修改"按钮,将当前记录显示到上方表单中,并将按钮改为"更新"按钮,如图 13-5 所示。

图 13-5

（3）用户登录后，单击"更新"按钮，即可更新博客信息。

5．搜索博客

（1）进入主页时，查询所有的博客信息。

（2）用户登录后，查询当前用户的全部博客记录。

（3）在主页顶部的搜索框中输入内容，可以搜索标题中包含该关键字的博客列表，如图 13-6 所示。

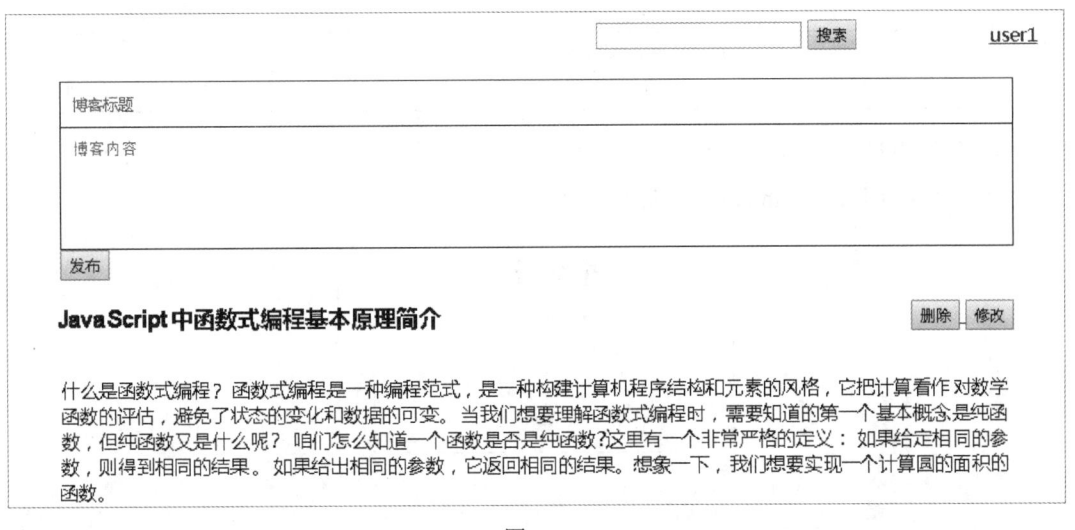

图 13-6

13.3 设计思路

1．工程设计

创建 Dynamic Web Project 项目——Blogs，集成 SSM 框架。

2．文件设计

文件设计如表 13-1 所示。

表 13-1

类型	文件/类	说明
Java 文件	com.blogs.mapper.UserMapper	User 数据库映射接口类
	com.blogs.mapper.BlogMapper	Blog 数据库映射接口类
	com.blogs.BlogMapper.xml	Blog 映射器 XML 文件
	com.blogs.controller.UserController	User 控制类文件
	com.blogs.controller.BlogController	Blog 控制类文件
	com.blogs.po.User	User 实体类
	com.blogs.po.Blog	Blog 实体类
JSP 文件	WEB-INF/jsp/login.jsp	登录页
	WEB-INF/jsp/index.jsp	主页
css 文件	css/blog.css	页面样式
js 文件	js/blog.js	页面脚本文件
配置文件	src/db.properties	数据库配置文件
	src/log4j.properties	日志配置文件
	src/applicationContext.xml	Spring 框架配置文件
	src/springmvc.xml	SpringMVC 框架配置文件

3. 页面设计

(1) 登录页面为 login.jsp, 如图 13-7 所示。

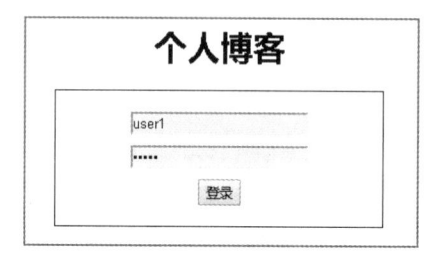

图 13-7

(2) 博客管理页面为 index.jsp, 如图 13-8 所示。

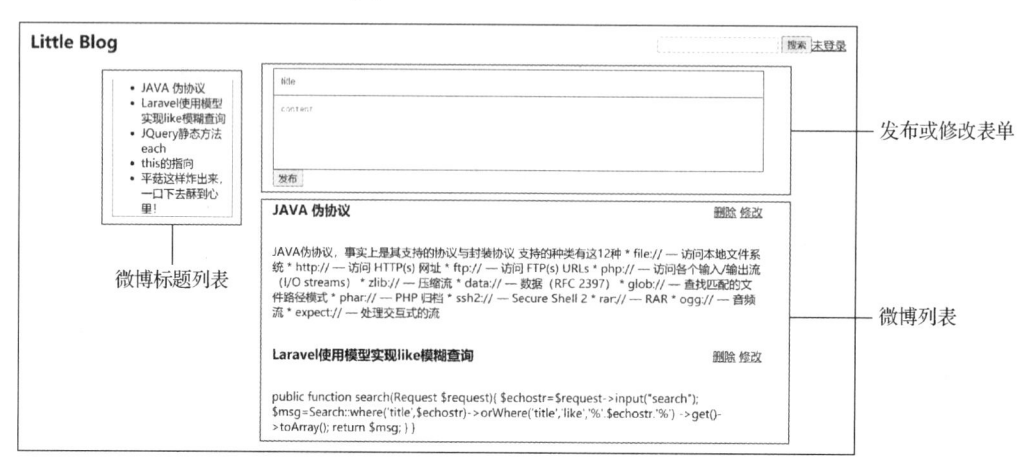

图 13-8

4．路由设计

（1）用户相关路由如表 13-2 所示。

表 13-2

路由	方法	响应
/login	GET	请求 login.jsp 页面 UserController::toLogin( )函数
/index 或 /	GET	请求 index.jsp 页面 BlogController::search( )函数
/user/login	POST	UserController::login( )函数
/user/logoff	GET	UserController::logoff( )函数

（2）博客相关路由如表 13-3 所示。

表 13-3

路由	方法	响应
/blog/add	POST	BlogController::add( )函数
/blog/search	GET	BlogController::search( )函数
/blog/del/{id}	GET	BlogController::delete( )函数
/blog/mod/{id}	GET	BlogController::toUpdate( )函数
/user/mod/	POST	BlogController::update( )函数

5．控制类

（1）用户控制类：UserController，使用@Controller 注解。
（2）博客控制类：BlogController，使用@Controller 注解。

6．模型类

（1）用户实体类：User。
（2）博客实体类：Blog。

7．创建数据库脚本文件 blogdb.sql

创建 blogdb 数据库，创建 blogs 和 users 表，代码如下。

```sql
drop database if exists blogdb;
create database blogdb;
use blogdb;

SET NAMES utf8mb4;
SET FOREIGN_KEY_CHECKS=0;

-- ----------------------------
-- Table structure for 'blogs'
-- ----------------------------
DROP TABLE IF EXISTS 'blogs';
CREATE TABLE 'blogs'(
  'id' int(11) NOT NULL AUTO_INCREMENT,
  'user_id' int(11) NOT NULL,
  'title' varchar(128) NOT NULL,
  'content' text DEFAULT NULL,
  'create_time' timestamp NULL DEFAULT current_timestamp() ON UPDATE
```

```sql
current_timestamp(),
  PRIMARY KEY ('id'),
  KEY 'user_id' ('user_id'),
  CONSTRAINT 'blogs_ibfk_1' FOREIGN KEY ('user_id') REFERENCES 'users' ('id')
) ENGINE=InnoDB AUTO_INCREMENT=2 DEFAULT CHARSET=utf8;

-- ----------------------------
-- Records of blogs
-- ----------------------------
INSERT INTO 'blogs' VALUES ('1', '1', 'this的指向', '通常函数中的this指向的是调用函数的对象(谁调用指向谁);\r\n事件函数中的this,通常指向的是绑定事件的事件源元素\r\n;构造函数中的this(使用new调用构造函数创建对象),通常指向的是new所创建出来的对象本身\r\n;全局范围的this,通常指向的是全局对象(浏览器中是window)', '2019-08-27 20:05:43');
INSERT INTO 'blogs' VALUES ('2', '1', 'JQuery静态方法each()', '1.首先是静态所以直接可以用[jQuery./$.+方法名]来调用;\r\n2.格式为$.each(object,[callback]) >>> callback中第一参数是index,第二参数是value;\r\n3.有返回值,返回值就是其遍历的数组;\r\n4.不可以在回调函数中对遍历数组进行处理;', '2019-08-27 20:06:19');
INSERT INTO 'blogs' VALUES ('3', '1', 'JAVA伪协议', 'JAVA伪协议,事实上是其支持的协议与封装协议\r\n支持的种类有12种 \r\n* file:// — 访问本地文件系统 \r\n* http:// — 访问 HTTP(s) 网址 \r\n* ftp:// — 访问 FTP(s) URLs \r\n* Java:// — 访问各个输入/输出流(I/O streams) \r\n* zlib:// — 压缩流 \r\n* data:// — 数据(RFC 2397) \r\n* glob:// — 查找匹配的文件路径模式 \r\n* phar:// — JAVA 归档 \r\n* ssh2:// — Secure Shell 2 \r\n* rar:// — RAR \r\n* ogg:// — 音频流 \r\n* expect:// — 处理交互式地流', '2019-08-27 20:08:07');
INSERT INTO 'blogs' VALUES ('4', '2', 'Laravel使用模型实现like模糊查询', 'public function search(Request $request){\r\n    $echostr=$request->input(\"search\");\r\n $msg=Search::where(\'title\',$echostr)->orWhere(\'title\',\'like\',\'%\'.$echostr.\'%\')\r\n ->get()->toArray();\r\n return $msg;\r\n    }\r\n}', '2019-08-27 20:45:58');

-- ----------------------------
-- Table structure for 'users'
-- ----------------------------
DROP TABLE IF EXISTS 'users';
CREATE TABLE 'users'(
  'id' int(11) NOT NULL AUTO_INCREMENT,
  'account' varchar(20) NOT NULL,
  'password' varchar(32) NOT NULL,
  PRIMARY KEY ('id')
) ENGINE=InnoDB AUTO_INCREMENT=2 DEFAULT CHARSET=utf8;

-- ----------------------------
-- Records of users
-- ----------------------------
INSERT INTO 'users' VALUES ('1', 'user', 'user');
```

```
INSERT INTO 'users' VALUES ('2', 'user1', 'user1');
```

13.4 实验实施（跟我做）

步骤 1：创建工程

（1）创建 Dynamic Web Project 工程 Blogs。

（2）工程目录如图 13-9 所示。

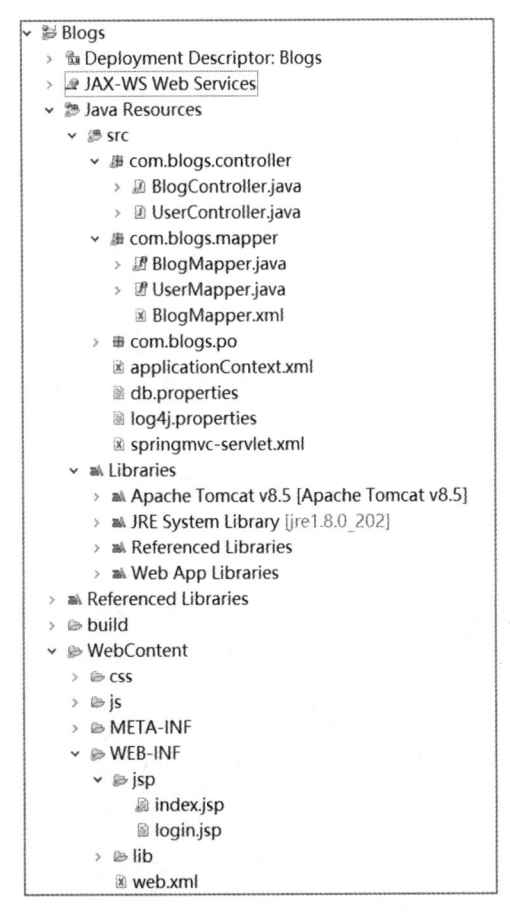

```
v ⊌ Blogs
  > ⌂ Deployment Descriptor: Blogs
  > ⌂ JAX-WS Web Services
  v ⌂ Java Resources
    v ⌂ src
      v ⌂ com.blogs.controller
        > ⌂ BlogController.java
        > ⌂ UserController.java
      v ⌂ com.blogs.mapper
        > ⌂ BlogMapper.java
        > ⌂ UserMapper.java
          ⌂ BlogMapper.xml
      ⌂ com.blogs.po
        ⌂ applicationContext.xml
        ⌂ db.properties
        ⌂ log4j.properties
        ⌂ springmvc-servlet.xml
    v ⌂ Libraries
      > ⌂ Apache Tomcat v8.5 [Apache Tomcat v8.5]
      > ⌂ JRE System Library [jre1.8.0_202]
      > ⌂ Referenced Libraries
      > ⌂ Web App Libraries
  > ⌂ Referenced Libraries
  > ⌂ build
  v ⌂ WebContent
    > ⌂ css
    > ⌂ js
    > ⌂ META-INF
    v ⌂ WEB-INF
      v ⌂ jsp
        ⌂ index.jsp
        ⌂ login.jsp
      > ⌂ lib
        ⌂ web.xml
```

图 13-9

（3）在 MySQL 中创建数据库。

启动 MySQL 命令行，输入"mysql -uroot –p"命令登录 MySQL 数据库，使用 source 命令导入数据库脚本 blogdb.sql，脚本内容参考设计思路中的创建数据库脚本文件。

（4）在 db.properties 配置数据库连接参数。

在 src 文件夹下的 db.properties 文件中配置 MySQL 用户名、密码和数据库名。

```
jdbc.driver=com.mysql.cj.jdbc.Driver
jdbc.url=jdbc:mysql://localhost:3306/blogdb?useUnicode=true&characterEnc
oding=utf8
jdbc.username=root
```

jdbc.password=123456

（5）配置 Tomcat 服务器，发布 Blogs 项目。

（6）在地址栏中输入网址 http://localhost:8080/Blogs，打开浏览器。

步骤 2：创建控制类

（1）在 UserController 类中创建 toLogin( )、login( )、logoff( )函数。

（2）在 BlogController 类中创建 search( )、toUpdate( )、add( )、delete( )、update( )函数。

步骤 3：配置路由

1. UserController 类

（1）进入登录页面（GET）。

```
@RequestMapping(value="/login", method=RequestMethod.GET)
public String toLogin(){
    return "login"; //返回 login.jsp 页面
}
```

（2）用户登录（POST）。

```
@RequestMapping(value="/user/login", method=RequestMethod.POST)
public String login(HttpSession session, String account, String password, Model model)
```

（3）用户退出（GET）。

```
@RequestMapping(value="/user/logoff", method=RequestMethod.GET)
public String logoff(HttpSession session)
```

2. BlogControleller 类

（1）进入主页，先查询博客获得数据，再显示主页（GET）。

```
@GetMapping(value={ "/","/index", "/blog/search" })
public String search(HttpSession session, HttpServletRequest request, Model model)
```

（2）添加博客，使用中间件验证用户是否登录（POST）。

```
@PostMapping(value="/blog/add")
public String add(HttpSession session, String title, String content)
```

（3）删除博客，使用中间件验证用户是否登录（GET）。

```
@GetMapping(value="/blog/del/{id}")
public String delete(@PathVariable int id)
```

（4）修改博客，使用中间件验证用户是否登录（GET）。

```
@GetMapping(value="/blog/mod/{id}")
public String toUpdate(@PathVariable int id, Model model)


@GetMapping(value="/blog/mod/{id}")
public String toUpdate(@PathVariable int id, Model model)
```

步骤 4：创建实体类

创建 com.blogs.po 包，在包中创建 User 类和 Blog 类。

1. 创建 User 类

```java
package com.blogs.po;

import org.apache.ibatis.type.Alias;

@Alias("User")
public class User{
  private Integer id;        //标识符
  private String account;    //账号
  private String password;   //密码

  public Integer getId(){
    return id;
  }
  public void setId(Integer id){
    this.id=id;
  }
  public String getAccount(){
    return account;
  }
  public void setAccount(String account){
    this.account=account;
  }
  public String getPassword(){
    return password;
  }
  public void setPassword(String password){
    this.password=password;
  }

}
```

2. 创建 Blog 类

```java
package com.blogs.po;

import java.util.Date;

import org.apache.ibatis.type.Alias;

@Alias("Blog")
public class Blog{
  private int id;          //标识符
  private User user;       //发布博客用户
```

```java
    private String title;      //博客标题
    private String content;    //博客内容
    private Date create_time;  //创建时间

    public int getId(){
        return id;
    }

    public void setId(int id){
        this.id=id;
    }

    public User getUser(){
        return user;
    }

    public void setUser(User user){
        this.user=user;
    }

    public String getTitle(){
        return title;
    }

    public void setTitle(String title){
        this.title=title;
    }

    public String getContent(){
        return content;
    }

    public void setContent(String content){
        this.content=content;
    }

    public Date getCreate_time(){
        return create_time;
    }

    public void setCreate_time(Date create_time){
        this.create_time=create_time;
    }

}
```

步骤5：实现登录功能

1. 编写 UserController::login( )方法

（1）从 request 中取得表单中的账号和密码。
（2）调用 where( )函数进行过滤，获得账号和密码相同的用户。
（3）若查询到用户，则表示登录成功。在 Session 中保存账号和用户的 id，跳转到首页。若登录失败，则返回登录页。

```java
package com.blogs.controller;

import javax.servlet.http.HttpSession;

import org.springframework.beans.factory.annotation.Autowired;
import org.springframework.stereotype.Controller;
import org.springframework.ui.Model;
import org.springframework.web.bind.annotation.RequestMapping;
import org.springframework.web.bind.annotation.RequestMethod;

import com.blogs.mapper.UserMapper;
import com.blogs.po.User;

@Controller
public class UserController{
  @Autowired
  private UserMapper userMapper=null;

  @RequestMapping(value="/login", method=RequestMethod.GET)
  public String toLogin(){
    return "login";
  }

  @RequestMapping(value="/user/login", method=RequestMethod.POST)
  public String login(HttpSession session, String account, String password, Model model){
    //获取账号和密码,构建User对象
    User user=new User();
    user.setAccount(account);
    user.setPassword(password);
    //调用 UserMapper 接口的 findUser 函数,查询出 User 对象
    user=userMapper.findUser(user);
    if(user!= null){
      //将用户的 id 和账号名保存到 Session 中
      session.setAttribute("uid", user.getId());
      session.setAttribute("account", user.getAccount());
      return "redirect:/index"; // 重定向到/index 路由
    } else{
```

```java
            model.addAttribute("account", account);
            model.addAttribute("msg", "用户名或者密码错误！请重新登录！");
            return "login";
        }
    }

    @RequestMapping(value="/user/logoff", method=RequestMethod.GET)
    public String logoff(HttpSession session){
        //从 Session 中移除用户信息
        session.removeAttribute("uid");
        session.removeAttribute("account");
        return "redirect:/index";
    }
}
```

2. 创建 login.jsp 文件

代码如下：

```jsp
<%@ page language="java" contentType="text/html; charset=UTF-8"
    pageEncoding="UTF-8"%>
<%
String basePath=request.getScheme()+"://"+request.getServerName()+":"+request.getServerPort()+request.getContextPath()+"/";
String message="";
if(request.getAttribute("message")!=null){

  message=(String) request.getAttribute("message");
}
%>
<!DOCTYPE html>
<html>
<head>
<base href="<%=basePath%>">
<meta charset="utf-8">
<link rel="stylesheet" href="css/blog.css">
<title>个人博客</title>
</head>
<body>
  <h1>个人博客</h1>
  <form class="log" action="user/login" method="post">
    <input type="text" name="account" id="account" placeholder="输入账号"/>
    <input type="password" name="password" id="password"
      placeholder="输入密码"/> <label><%=message%></label>
       <input type="submit" value="登录"/>
  </form>
</body>
</html>
```

3. 创建 index.jsp 文件

代码如下:

```html
<!DOCTYPE html>
<html>
    <head>
    <meta charset="utf-8"/>
    <link rel="stylesheet" href="css/blog.css">
    <title>个人博客</title>
    </head>
    <body>
        <header>
            <nav>
                <h2>Little Blog</h2>
                <p>
                <%
                if(account==""){
                %>
                <a href="login">未登录</a>
                <%
                } else{
                %>
                <a href="user/logoff">${account}</a>
                <%
                }
                %>
                </p>
            </nav>
        </header>
    </body>
</html>
```

步骤 6: 显示博客列表

1. 创建页面样式文件

在 public/css 文件夹中创建 blog.css 文件。

```css
body{ min-width:677px}
/* 登录页 */
h1{text-align: center;}
.log input{
    display: block;
    margin:10px auto;
}
form.log{
    width: 300px;
    padding: 10px;
    margin: auto;
```

```css
    border: 1px black solid;
}
/* 首页 */
nav{
    display: flex;
    align-items: center;
}
nav h2{flex: 1;}
nav form{
    flex: 3;
    text-align: right;
}
nav #account{
    flex: 1;
    text-align: center;
}
aside{
    float: left;
    width: 15%;
    margin-left: 10%;
}
aside ul{ border:1px black solid;}
section{
    float: right;
    width: 60%;
    margin-right: 10%;
}
section form div{
    border: 1px black solid;
}
section form input[type="text"]{
    display: block;
    width: 100%;
    border: 0;
    border-bottom: 1px black solid;
}
section form textarea{
    border: 0;
    width: 100%;
    resize: none;
}
section input[type="text"],textarea{
    box-sizing: border-box;
    outline: 0;
    padding: 10px;
}
article .b-t{
```

```
    display: flex;
    align-items: center;
    justify-content: space-between;
}
```

2. 在 index.jsp 文件中添加搜索框

代码如下：

```html
<!DOCTYPE html>
<html>
    <head>
    <meta charset="utf-8"/>
    <link rel="stylesheet" href="css/blog.css">
    <title>个人博客</title>
    </head>
    <body>
        <header>
            <nav>
                <h2>Little Blog</h2>
                <form action="/blog/search" method="get">
                    <input type="text" name="keyword"/>
                    <input type="submit" value="搜索"/>
                </form>
                ……
            </nav>
        </header>
    </body>
</html>
```

3. 编写 BlogController::search( )方法

从 request 中获得 keyword 属性，从 Session 中获得 userid 值。根据 keyword 与 userid 的情况，查询方式可分为以下 4 种情况。

（1）未登录，未搜索。

（2）未登录，搜索。

（3）登录，未搜索。

（4）登录，搜索。

创建 com.blog.controller 包，在该包中创建 Blog Controller 类，添加控制类代码，代码如下：

```java
@Controller
public class BlogController{
    @Autowired
    private BlogMapper blogMapper=null;

    @GetMapping(value={"/","/index", "/blog/search"})
    public String search(HttpSession session, HttpServletRequest request, Model model){
        //获得查询关键字
```

```
        String keyword=request.getParameter("keyword");

        //判断用户是否登录
        if(session.getAttribute("uid")!=null){//用户已登录
          int userId=(int)session.getAttribute("uid");
          if(keyword!=null){
            //通过关键词查询登录用户的博客
            Map<String, Object> params=new HashMap();
            params.put("uid", userId);
            params.put("keyword", '%'+keyword+'%');
            List<Blog>blogs=blogMapper.search(params);
            model.addAttribute("blogs", blogs);
          } else{
            //查询用户发布的博客
            List<Blog>blogs=blogMapper.getBlogsByUserid(userId);
            model.addAttribute("blogs", blogs);
          }
        } else{//用户未登录
          if(keyword!=null){
            //通过关键词查询博客
            List<Blog>blogs=blogMapper.getBlogsByKeyword(keyword);
            model.addAttribute("blogs", blogs);
          } else{
            //查询所有博客
            List<Blog>blogs=blogMapper.getBlogs();
            model.addAttribute("blogs", blogs);
          }
        }
        return "index";
      }

}
```

4. 在 index.jsp 文件中显示查询的记录

（1）添加<aside>侧边栏，使用 for 循环遍历 blogs，显示博客标题的列表。使用<ul><li></li></ul>标签。

（2）使用 if 判断 blogs 变量是否存在，若存在则显示博客标题列表，若不存在则不显示。

```
……
  <body>
    <header>
    ……
    </header>
    <aside>
    <%
    if(blogs!=null){
    %>
```

```
    <ul>
     <%
     for(Blog b:blogs){
     %>
     <li><%=b.getTitle()%></li>
     <%
     }
     %>
    </ul>
    <%
    }
    %>
   </aside>
  </body>
</html>
```

（3）定义<section>标签显示博客详细列表，每篇博客都使用<article>标签。

（4）分别显示博客标题和内容。

```
……
<body>
……
  <section>
    <%
    if(blogs!=null){
      for(Blog b:blogs){
    %>
    <article>
      <div class="b-t">
        <h3><%=b.getTitle()%></h3>
        <div class="act">
          <a href="blog/del/<%=b.getId()%>">删除</a>
          <a href="blog/mod/<%=b.getId()%>">修改</a>
        </div>
      </div>
      <p id="b-c">
        <%=b.getContent()%>
      </p>
    </article>
    <%
      }
    }
    %>
  </section>
```

步骤 7：添加博客

（1）在 index.jsp 文件中添加发布博客的表单。

```
<section>
```

```html
<form action="blog/add" method="post">
  <div>
    <input type="text" id="title" name="title" placeholder="title"/>
    <textarea rows="5" id="content" name="content" placeholder="content"></textarea>
  </div>
  <input type="submit" value="发布"/>
</form>
```

（2）编写 BlogController::add( )函数。

从 request 对象中获得表单中的 title 和 content 值。

创建 Blog 对象，设置字段值，调用 BlogMapper.insertOne( )函数保存数据，返回首页。

```java
@PostMapping(value="/blog/add")
public function add(Request $request){
    //获得当前登录的用户
    User user=new User();
    if(session.getAttribute("uid")!=null){
      int uid=(int)session.getAttribute("uid");
      user.setId(uid);
    }
    //创建Blog对象
    Blog blog=new Blog();
    blog.setTitle(title);
    blog.setContent(content);
    blog.setUser(user);
    try{
      blogMapper.insertOne(blog); //插入博客
      return "redirect:/index";
    } catch (Exception e){
      return "redirect:/login";
    }
}
```

步骤 8：修改和删除博客

（1）添加"删除"和"修改"按钮，给按钮添加<a>标签。

删除：href="/blog/del/博客 id"。

修改：href="/blog/mod/博客 id"。

```jsp
……
<section>
<%
if(blogs!=null){
  for(Blog b:blogs){
%>
<article>
  <div class="b-t">
    <h3><%=b.getTitle()%></h3>
```

```html
    <div class="act">
      <a href="blog/del/<%=b.getId()%>">删除</a> <a
      href="blog/mod/<%=b.getId()%>">修改</a>
    </div>
  </div>
  <p id="b-c">
    <%=b.getContent()%>
  </p>
</article>
<%
  }
}
%>
......
```

（2）编写 BlogController::delete( )函数。

```java
@GetMapping(value="/blog/del/{id}")
public String delete(@PathVariable int id){
  try{
    blogMapper.deleteById(id); //通过id删除博客
    return "redirect:/index";
  }catch (Exception e){
    return "redirect:/login";
  }
}
```

（3）编写 BlogController::update( )函数。

```java
@PostMapping(value="/blog/mod")
  public String update(HttpSession session, int id, String title, String content){
    //获得当前登录的User对象
    User user=new User();
    if(session.getAttribute("uid")!=null){
      int uid=(int) session.getAttribute("uid");
      user.setId(uid);
    }
    //构建Blog对象
    Blog blog=new Blog();
    blog.setId(id);
    blog.setTitle(title);
    blog.setContent(content);
    blog.setUser(user);
    try{
      blogMapper.updateOne(blog); //更新博客信息
      return "redirect:/index";
    } catch (Exception e){
      return "redirect:/login";
    }
  }
```

（4）编写 BlogController::toUpdate( )函数。

```
@GetMapping(value="/blog/mod/{id}")
public String toUpdate(@PathVariable int id, Model model){
  try{
    Blog blog=blogMapper.getBlog(id); //通过id查询博客
    model.addAttribute("blog", blog);
    return "index";
  } catch (Exception e){
    return "redirect:/login";
  }
}
```

（5）修改 index.jsp 文件中的发布博客表单。

```
<form action="blog/mod" method="post">
  <div>
    <input type="hidden" id="id" name="id" value="<%=blog.getId()%>"/>
    <input type="text" id="title" name="title" placeholder="title"
      value="<%=blog.getTitle()%>"/>
    <textarea id="a-c" rows="5" id="content" name="content"
      placeholder="content"><%=blog.getContent()%></textarea>
  </div>
  <input type="submit" value="更新"/>
</form>
```

第 14 章
RESTful API 规范：视频列表

14.1 实验目标

（1）理解 RESTful API 设计规则。
（2）理解使用 SpringMVC 框架搭建 RESTful API 接口的方法。
（3）通过 RESTful API，应用 AJAX 和 Java 动态网站进行前后端数据交互。
（4）综合运用 RESTful API 和 Java Web 编程，开发"视频列表"程序。

14.2 实验任务

实现视频列表动态页面功能：页面包括搜索框、视频列表、分页栏 3 个部分，单击页面搜索栏中的"搜索"按钮或单击分页栏中的分页按钮，页面向服务器发送请求获取视频数据，将获取的数据显示到页面的视频列表中。

（1）使用 RESTful 格式设计视频列表 API。

API 格式："域名/路径/资源?筛选条件"。

API 作用：请求 API 后能返回视频列表数据。

硬编码数据如下：

```
{\"name\":\"视频 1\",\"v_src\":\"video/coffee.mp4\"},
{\"name\":\"视频 2\",\"v_src\":\"video/coffee.mp4\"},
{\"name\":\"视频 3\",\"v_src\":\"video/coffee.mp4\"},
{\"name\":\"视频 4\",\"v_src\":\"video/coffee.mp4\"},
{\"name\":\"视频 5\",\"v_src\":\"video/coffee.mp4\"},
{\"name\":\"视频 6\",\"v_src\":\"video/coffee.mp4\"}
```

（2）搭建 Java 动态网站，集成 Spring+SpringMVC 框架，定义视频列表 API 接口。
（3）创建 videoList.html 页面，使用 AJAX 请求视频列表 API 接口，通过访问接口获取视频列表数据，返回数据格式如下：

```
{
  \"code\":200,
```

```
\"data\":[{\"name\":\"视频1\",\"v_src\":\"video/coffee.mp4\"}
         {\"name\":\"视频2\",\"v_src\":\"video/coffee.mp4\"},
         {\"name\":\"视频3\",\"v_src\":\"video/coffee.mp4\"},
         {\"name\":\"视频4\",\"v_src\":\"video/coffee.mp4\"},
         {\"name\":\"视频5\",\"v_src\":\"video/coffee.mp4\"},
         {\"name\":\"视频6\",\"v_src\":\"video/coffee.mp4\"}]
}
```

将视频列表数据显示在 videoList.html 页面上，如图 14-1 所示。

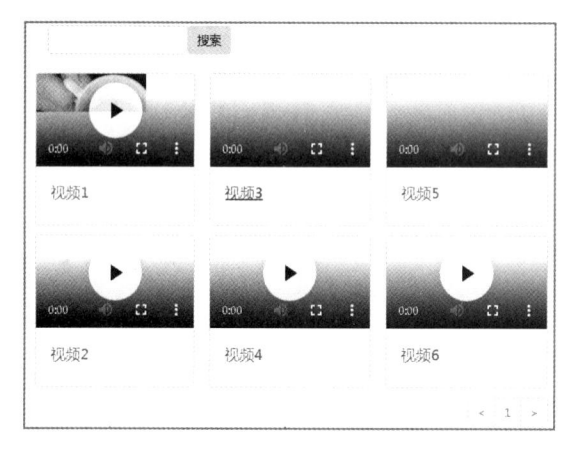

图 14-1

14.3 设计思路

1. 工程设计

创建 Dynamic Web Project 项目——Rest，集成 Spring 和 SpringMVC 框架。

2. 文件设计

文件设计如表 14-1 所示。

表 14-1

类型	文件	说明
html 文件	videoList.html	视频列表页面
css 文件	css/bootstrap.min.css	bootstrap css 文件
js 文件	js/search.js	发送 AJAX 请求，获取数据，页面局部刷新
	js/jquery-3.2.1.min.	jQuery 库
java 文件	com.rest.VideoController	视频控制器类
	com.rest.Video	视频模型类
	com.rest.ResultModel	返回结果模型对象
配置文件	src/springmvc.xml	SpringMVC 框架配置文件
	src/applicationContext.xml	Spring 框架配置文件
	WEB-INF/web.xml	网站主配置文件

3．设计视频列表 API

（1）API 中涉及动词 POST、GET、DELETE、PUT。
（2）返回值为{\"code\":200,\"data\":[ ]}。
（3）基本格式为 http://域名/路径/资源?过滤参数，如图 14-2 所示。

```
        域名          资源         过滤参数
http://localhost/videos/list{?Ppage_no, keyword}
```

图 14-2

4．设计视频列表接口

（1）创建 VideoController 类。

使用@RestController 注解，定义 ResultModel 类返回，添加 jackson 包。框架会将对象转为 JSON 格式的字符串返回。

在类上使用@RequestMapping(value="/api/videos")定义 API 的统一头部，表示该控制器类返回的资源为 videos。

使用@GetMapping("/list")注解定义路由，响应函数为 list 函数。

```
public ResultModel list()
```

（2）视频数据。

使用硬编码，返回视频列表数据，返回的 JSON 数据如下：

```
{
 \"code\":200,
 \"data\":[
   {\"name\":\"视频 1\",\"v_src\":\"video/coffee.mp4\"},
   {\"name\":\"视频 2\",\"v_src\":\"video/coffee.mp4\"},
   {\"name\":\"视频 3\",\"v_src\":\"video/coffee.mp4\"},
   {\"name\":\"视频 4\",\"v_src\":\"video/coffee.mp4\"},
   {\"name\":\"视频 5\",\"v_src\":\"video/coffee.mp4\"},
   {\"name\":\"视频 6\",\"v_src\":\"video/coffee.mp4\"}
 ]
}
```

5．设计视频列表页面 videoList.html

（1）创建 videoList.html 页面，页面包括 3 个部分：搜索框、视频列表（显示 6 条视频数据）、分页栏，如图 14-3 所示。

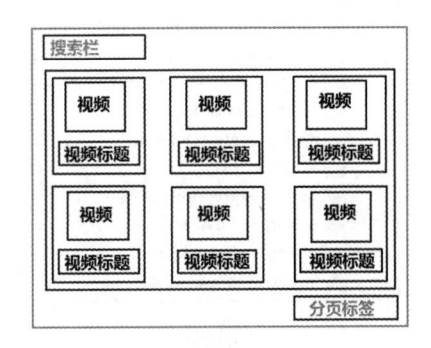

图 14-3

（2）定义一个 search( )函数，处理搜索按钮 onclick 单击事件和分页按钮 onclick 单击事件。

（3）在 search( )函数中使用 AJAX 调用视频列表接口，获取数据，将获取的数据显示到页面的视频列表中。

14.4 实验实施（跟我做）

步骤 1：创建工程

启动 Eclipse，选择"File"→"New"→"Dynamic Web Project"选项，在打开的向导窗口的"Project name"文本框中输入"Rest"工程名，在"Target runtime"下拉列表中选择"Apache Tomcat v8.5"选项。

单击"Next"按钮，在打开的"Web Module"配置窗口中选中"Generate web.xml deployment descriptor"复选框，向导将自动创建 web.xml 文件。

单击"Finish"按钮，完成创建工程的操作。在 Package Explorer 中可以看到新创建的 Rest 工程，如图 14-4 所示。

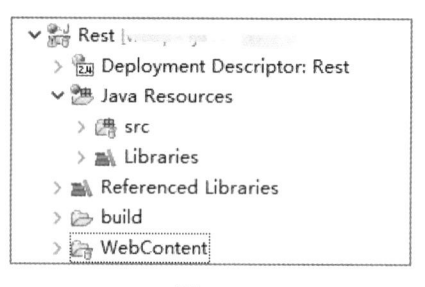

图 14-4

导入 Spring 和 SpringMVC 框架的包，以及依赖 commons-logging-1.2.jar 包。添加 jackson-annotations、jackson-core、jackson-databind 这 3 个 jackson 包，用来将对象转为 JSON 字符串，如图 14-5 所示。

图 14-5

在"WebContent"目录下添加如表 14-2 所示的目录。

表 14-2

文件夹	说明
/css/	存放网页中的 CSS 文件
/js/	存放网页中的 JS 文件
/video/	存放视频文件

将视频文件 coffee.mp4 放到 video 文件夹中。将 bootstrap.min.css 和 bootstrap-grid.min.css 文件放到 css 文件夹中。将 jquery-3.2.1.min.js 文件放到 js 目录中，并在其下创建 search.js 文件，如图 14-6 所示。

图 14-6

步骤 2：编写配置文件

在 src 下创建 applicationContext.xml 文件，编写代码如下。

```xml
<?xml version="1.0" encoding="UTF-8"?>
<beans xmlns="http://www.springframework.org/schema/beans"
  xmlns:context="http://www.springframework.org/schema/context"
  xmlns:xsi="http://www.w3.org/2001/XMLSchema-instance"
  xmlns:aop="http://www.springframework.org/schema/aop"
  xmlns:tx="http://www.springframework.org/schema/tx"
  xmlns:p="http://www.springframework.org/schema/p"
  xsi:schemaLocation="
    http://www.springframework.org/schema/context
    http://www.springframework.org/schema/context/spring-context.xsd
    http://www.springframework.org/schema/beans
    http://www.springframework.org/schema/beans/spring-beans.xsd
    http://www.springframework.org/schema/tx
    http://www.springframework.org/schema/tx/spring-tx.xsd
    http://www.springframework.org/schema/aop
    http://www.springframework.org/schema/aop/spring-aop.xsd">

</beans>
```

在 src 下创建 springmvc.xml 文件，编写代码如下。

```xml
<?xml version="1.0" encoding="UTF-8"?>
<beans xmlns="http://www.springframework.org/schema/beans"
  xmlns:xsi="http://www.w3.org/2001/XMLSchema-instance"
```

```xml
  xmlns:tx="http://www.springframework.org/schema/tx"
  xmlns:context="http://www.springframework.org/schema/context"
  xmlns:mvc="http://www.springframework.org/schema/mvc"
  xsi:schemaLocation="http://www.springframework.org/schema/beans
    http://www.springframework.org/schema/beans/spring-beans-3.0.xsd
    http://www.springframework.org/schema/tx
    http://www.springframework.org/schema/tx/spring-tx-3.0.xsd
    http://www.springframework.org/schema/context
    http://www.springframework.org/schema/context/spring-context-3.0.xsd
    http://www.springframework.org/schema/mvc
    http://www.springframework.org/schema/mvc/spring-mvc-3.0.xsd">

    <mvc:annotation-driven/>

    <!-- 处理器映射器 -->
    <bean
      class="org.springframework.web.servlet.handler.BeanNameUrlHandlerMapping"/>

    <!-- 处理器适配器 -->
    <bean
      class="org.springframework.web.servlet.mvc.SimpleControllerHandlerAdapter"/>

    <!-- 使用组件扫描 -->
    <context:component-scan base-package="com.rest"/>

    <!-- 配置 SpringMVC 视图解析器 -->
    <bean
      class="org.springframework.web.servlet.view.InternalResourceViewResolver">
      <!-- 路径前缀 -->
      <property name="prefix" value="/WEB-INF/jsp/"/>
      <!-- 路径后缀 -->
      <property name="suffix" value=".jsp"/>
      <!-- 前缀+视图逻辑名+后缀=真实路径 -->
    </bean>

</beans>
```

编写 web.xml 文件，代码如下。

```xml
<?xml version="1.0" encoding="UTF-8"?>
<web-app version="2.4" xmlns="http://JAVA.sun.com/xml/ns/j2ee"
  xmlns:xsi="http://www.w3.org/2001/xmlSchema-instance"
  xsi:schemaLocation="http://JAVA.sun.com/xml/ns/j2ee
     http://JAVA.sun.com/xml/ns/j2ee/web-app_2_4.xsd">

  <display-name>Rest</display-name>

  <listener>
    <listener-class>org.springframework.web.context.request.RequestContextListener
```

```xml
</listener-class>
    </listener>

    <!-- 加载 spring 容器 -->
    <context-param>
      <param-name>contextConfigLocation</param-name>
      <param-value>classpath:applicationContext.xml</param-value>
    </context-param>
    <listener>
      <listener-class>org.springframework.web.context.ContextLoaderListener</listener-class>
    </listener>

    <!-- 配置 DispatchcerServlet -->
    <servlet>
      <servlet-name>dispatcher</servlet-name>
      <servlet-class>org.springframework.web.servlet.DispatcherServlet</servlet-class>
        <init-param>
          <param-name>contextConfigLocation</param-name>
          <param-value>classpath:springmvc.xml</param-value>
        </init-param>
        <load-on-startup>1</load-on-startup>
    </servlet>
    <servlet-mapping>
      <servlet-name>dispatcher</servlet-name>
      <url-pattern>/</url-pattern>
    </servlet-mapping>

    <!-- 配置过滤器 将 POST 请求转换为 PUT 和 DELETE 请求 -->
    <filter>
      <filter-name>HiddenHttpMethodFilter</filter-name>
      <filter-class>org.springframework.web.filter.HiddenHttpMethodFilter</filter-class>
    </filter>
    <filter-mapping>
      <filter-name>HiddenHttpMethodFilter</filter-name>
      <url-pattern>/*</url-pattern>
    </filter-mapping>

    <!-- spring 字符编码过滤器 start -->
    <filter>
      <!--① Spring 编码过滤器 -->
      <filter-name>encodingFilter</filter-name>
      <filter-class>org.springframework.web.filter.CharacterEncodingFilter</filter-class>
        <!--② 编码方式 -->
```

```xml
    <init-param>
      <param-name>encoding</param-name>
      <param-value>UTF-8</param-value>
    </init-param>
    <!--③ 强制进行编码转换 -->
    <init-param>
      <param-name>forceEncoding</param-name>
      <param-value>true</param-value>
    </init-param>
  </filter>
  <!-- ② 过滤器的匹配 URL -->
  <filter-mapping>
    <filter-name>encodingFilter</filter-name>
    <url-pattern>/*</url-pattern>
  </filter-mapping>
  <!-- spring 字符编码过滤器 end -->

  <!-- 指明对于如上资源文件不采用 spring 的过滤器 -->
  <servlet-mapping>
    <servlet-name>default</servlet-name>
    <url-pattern>*.css</url-pattern>
  </servlet-mapping>
  <servlet-mapping>
    <servlet-name>default</servlet-name>
    <url-pattern>*.gif</url-pattern>
  </servlet-mapping>
  <servlet-mapping>
    <servlet-name>default</servlet-name>
    <url-pattern>*.jpg</url-pattern>
  </servlet-mapping>
  <servlet-mapping>
    <servlet-name>default</servlet-name>
    <url-pattern>*.js</url-pattern>
  </servlet-mapping>
  <servlet-mapping>
    <servlet-name>default</servlet-name>
    <url-pattern>*.mp4</url-pattern>
  </servlet-mapping>
  <servlet-mapping>
    <servlet-name>default</servlet-name>
    <url-pattern>*.html</url-pattern>
  </servlet-mapping>
</web-app>
```

步骤 3：实现 API 接口

1. 定义数据类 Video

创建 com.rest 包，创建 Video 类。由于在约定返回的 JSON 数据中，视频 URL 的 key

值为"v_src",所以要使用@JsonProperty 注解,定义转换时使用的 key 值,不设置时默认 key 值与属性名相同。

```java
package com.rest;

import com.fasterxml.jackson.annotation.JsonProperty;

public class Video{
  private String name;
  @JsonProperty(value="v_src")
  private String uri;

  public Video(String name, String uri){
    super();
    this.name=name;
    this.uri=uri;
  }

  public String getName(){
    return name;
  }

  public void setName(String name){
    this.name=name;
  }

  public String getUri(){
    return uri;
  }

  public void setUri(String uri){
    this.uri=uri;
  }

}
```

2. 定义返回的 ResultModel 模型类

在 com.rest 包中创建 ResultModel 类。

```java
package com.rest;

public class ResultModel{
  private Integer code;
  private Object data;

  public Integer getCode(){
    return code;
  }
```

```java
    public void setCode(Integer code){
        this.code=code;
    }

    public Object getData(){
        return data;
    }

    public void setData(Object data){
        this.data=data;
    }
}
```

3. 编写 VideoController 控制器类

在 com.rest 包中创建 VideoController 类，使用@RestController 类进行注解。

```java
package com.rest;

import java.util.ArrayList;
import java.util.List;

import org.springframework.web.bind.annotation.GetMapping;
import org.springframework.web.bind.annotation.RequestMapping;
import org.springframework.web.bind.annotation.RestController;

@RestController
@RequestMapping(value="/api/videos")
public class VideoController{

    @GetMapping("/list")
    public ResultModel list(){
        List<Video> videos=new ArrayList<Video>();
        videos.add(new Video("视频1", "video/coffee.mp4"));
        videos.add(new Video("视频2", "video/coffee.mp4"));
        videos.add(new Video("视频3", "video/coffee.mp4"));
        videos.add(new Video("视频4", "video/coffee.mp4"));
        videos.add(new Video("视频5", "video/coffee.mp4"));
        videos.add(new Video("视频6", "video/coffee.mp4"));

        ResultModel result=new ResultModel();
        result.setCode(200);
        result.setData(videos);
        return result;
    }
}
```

步骤 4：使用 RESTful API

（1）创建 videoList.html 页面，引入 bootstrap-grid.min.css 样式，页面包括搜索框、视频列表和分页栏，代码如下：

```html
<!DOCTYPE html>
<head>
  <meta charset="utf-8">
  <link rel="stylesheet" type="text/css" href="css/bootstrap.min.css"/>
  <script src="js/jquery-3.2.1.min.js"></script>
  <script src="js/search.js"></script>
  <title>视频列表</title>
</head>

<body>
  <div class="container">
    <nav class="navbar">
      <form class="form-inline mr-auto">
        <input id="keyword" type="text" class="form-control"/>
        <input class="btn" type="submit" value="搜索" onclick="return search();"/>
      </form>
    </nav>
    <div class="row">
      <div class="col-12">
        <div id="list" class="card-columns"></div>
      </div>
    </div>
  </div>
  <body>
```

页面效果如图 14-7 所示。

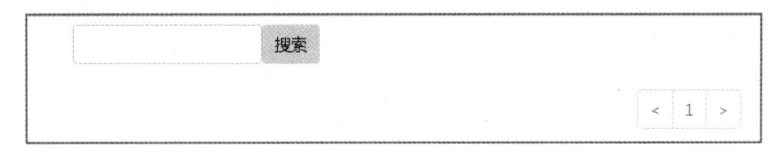

图 14-7

（2）在 js 文件夹中创建 search.js 文件，在 videoList.html 文件中导入。

编写 search( )函数。使用 AJAX 发送请求"/api/videos/list"获取视频列表数据，将其显示到界面上。

```javascript
$(function() {
  search();
});

//视频音乐
function search(){
```

```
//调用AJAX,请求视频列表资源: /api/videos/list
$.ajax({
  type: "get",
  url: "http://127.0.0.1:8080/api/videos/list",
  dataType: "json",
  headers:{
    Accept: "application/json; charset=utf-8"
  },
  success: function(result){
    //判断返回状态码
    if(result.code==200){
      //插入返回的视频
      $.each(result.data, function(key, value){
        var text="<div class='card'>"+
            "<a href='#'>"+
              "<video controls='controls' class='card-img-top' src='"+value.v_src+"' ></video>"+
              "<div class='card-body'><h5 class='card-title'>"+value.name+"</h5></div>"+
            "</a>"+
          "</div>";
        $("#list").append(text);

      });
    }
  },
  error: function(jqXHR, textStatus, errorThrown){
    alert(jqXHR.status);
    alert("error"+jqXHR.responseText);
  }
});
}
```

效果如图 14-1 所示。

第 15 章 AJAX：天气预报

15.1 实验目标

（1）掌握 XMLHttpRequest 对象的创建和使用方法。
（2）掌握使用 AJAX 服务器发送异步请求的方法。
（3）掌握 AJAX 服务器响应的方法。
（4）掌握 JSON 数据格式解析的方法。
（5）综合运用 AJAX 技术，实现"天气预报"移动端程序。

15.2 实验任务

创建天气预报页面，适配移动端访问，使用 AJAX 请求 Java，获取北京、上海、广州、深圳和武汉这 5 个城市的天气数据，每次请求 Java 都会随机生成天气数据。
天气数据内容如下：

{"name" : "北京","min": "20℃","max" :"20℃","weather" : "多云转阴"}

然后使用 JavaScript 操作 DOM 将获取的天气信息实时更新至页面。页面效果如图 15-1 所示。

图 15-1

15.3 设计思路

1．工程设计

创建 Dynamic Web Project 项目——Weather。

2．文件设计

文件设计如表 15-1 所示。

表 15-1

类型	文件/类	说明
html 文件	index.html	天气预报首页 html 页面，展示天气信息
Java 文件	com.weather.servlet.WeatherServlet	返回 JSON 格式的天气预报文件
	com.weather.model.WeatherItem	天气项数据类

3．类设计

创建 WeatherServlet 类，接收天气数据请求，返回 JSON 格式的天气数据。

city 变量：用于存放城市数据。

使用 ArrayList 类创建天气列表，以硬编码方式输入天气数据。

使用 fastjson 对变量进行 JSON 编码。

使用 PrintWriter 输出 JSON。

4．页面设计

（1）设计移动端的天气预报页面展示天气信息，使用语义化标签搭建页面结构，如图 15-2 所示。

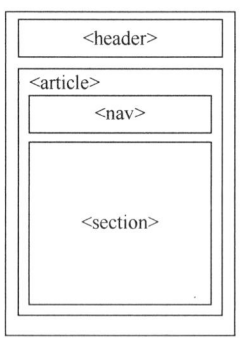

图 15-2

（2）设计天气预报页面样式。

使用 flex 弹性布局设置城市导航栏在一行中显示。

使用 transition 过渡属性给当前被单击的城市添加变宽的效果。

使用 rem、em、百分比（%）单位设置元素大小。

5．数据交互

（1）通过 AJAX 获取 JSON 格式的天气数据，实现异步获取数据。

使用 open( )方法创建请求，设置请求类型、请求路径、异步请求。

使用 send( )方法发起请求。

为 onreadystatechange 属性设置函数，监听请求状态。

使用 responseText( ) 方法获取返回的数据，使用 JSON.parse( ) 方法将字符串解析成 JSON 格式。

（2）通过 JavaScript 操作 DOM 实现天气信息实时更新至页面，实现异步刷新。

15.4 实验实施（跟我做）

步骤 1：创建项目和文件

（1）创建项目：项目名为 weather。

（2）创建页面。

index.html：天气预报首页。

（3）创建类。

WeatherServlet：返回 JSON 格式的天气预报。

WeatherItem：天气项数据类。

如图 15-3 所示。

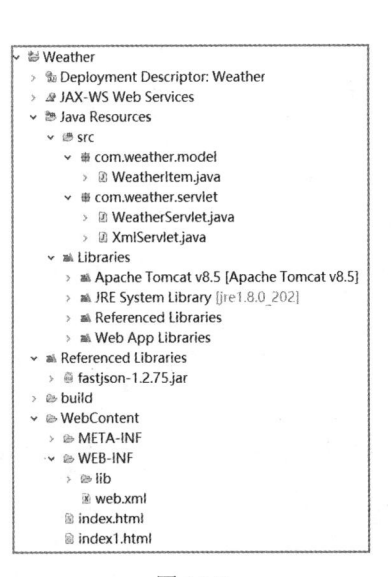

图 15-3

步骤 2：实现 Servlet 类

（1）添加 fastjson.jar 包。

（2）创建 WeatherItem 类。

```
package com.weather.model;
import com.alibaba.fastjson.annotation.JSONField;

public class WeatherItem{
    // 设置转成 JSON 后的 key 值
    @JSONField(name="name")
```

```java
    private String name;           //城市名称
@JSONField(name="min")
    private int min;               //最低温度
@JSONField(name="max")
    private int max;               //最高温度
@JSONField(name="weather")
    private String weather;        //天气描述

public WeatherItem(String name, int min, int max, String weather){
    super();
    this.name=name;
    this.min=min;
    this.max=max;
    this.weather=weather;
}

public String getName(){
    return name;
}

public void setName(String name){
    this.name=name;
}

public String getMin(){
    return min+"℃";
}

public void setMin(int min){
    this.min=min;
}

public String getMax(){
    return max+"℃";
}

public void setMax(int max){
    this.max=max;
}

public String getWeather(){
    return weather;
}

public void setWeather(String weather){
    this.weather=weather;
}
```

}

（3）创建 WeatherServlet 类。

```java
package com.weather.servlet;

import java.io.IOException;
import java.io.PrintWriter;
import java.util.ArrayList;
import java.util.List;
import java.util.Random;

import javax.servlet.ServletException;
import javax.servlet.annotation.WebServlet;
import javax.servlet.http.HttpServlet;
import javax.servlet.http.HttpServletRequest;
import javax.servlet.http.HttpServletResponse;

import com.alibaba.fastjson.JSON;
import com.weather.model.WeatherItem;

/**
 * Servlet implementation class WeatherServlet
 */
@WebServlet("/ListWeather")
public class WeatherServlet extends HttpServlet{
  private static final long serialVersionUID=1L;

  /**
   * @see HttpServlet#HttpServlet()
   */
  public WeatherServlet(){
    super();
    // TODO Auto-generated constructor stub
  }

  /**
   * @see HttpServlet#doGet(HttpServletRequest request, HttpServletResponse response)
   */
  protected void doGet(HttpServletRequest request, HttpServletResponse response) throws ServletException, IOException{
    //初始化天气数据列表
    List<WeatherItem>data=new ArrayList<WeatherItem>();
    Random rand=new Random();
    data.add(new WeatherItem("北京", rand.nextInt(21), rand.nextInt(21)+20, "多云转阴"));
    data.add(new WeatherItem("上海", rand.nextInt(21), rand.nextInt(21)+20,
```

```
"晴"));
        data.add(new WeatherItem("广州", rand.nextInt(21), rand.nextInt(21)+20,
"小雨转晴"));
        data.add(new WeatherItem("深圳", rand.nextInt(21), rand.nextInt(21)+20,
"晴"));
        data.add(new WeatherItem("武汉", rand.nextInt(21), rand.nextInt(21)+20,
"晴"));

        //获得城市天气信息
        String city=request.getParameter("city");
        String result="";
        if(city.equals("北京")){
          result=JSON.toJSONString(data.get(0));
        //将 WeatherItem 类对象转为 JSON 字符串
        }else if(city.equals("上海")){
          result=JSON.toJSONString(data.get(1));
        }else if(city.equals("广州")){
          result=JSON.toJSONString(data.get(2));
        }else if(city.equals("深圳")){
          result=JSON.toJSONString(data.get(3));
        }else{
          result=JSON.toJSONString(data.get(4));
        }

        //返回 JSON 字符串
        response.setCharacterEncoding("UTF-8");
        response.setContentType("application/json; charset=utf-8");
        PrintWriter writer=response.getWriter();
        writer.append(result);
    }
}
```

步骤 3：制作 HTML 页面

（1）在 index.html 页面的<head>标签中设置移动端适配属性 viewport，使用语义化标签搭建页面结构，代码如下：

```
<!DOCTYPE html>
<html>
    <head>
        <meta charset="utf-8">
        <meta name='viewport' content='width=device-width, initial-scale=1.0'>
        <title>天气预报</title>
    </head>
    <body>
        <header>
            <h4>天气预报</h4>
        </header>
```

```html
    <nav class="btn">
        <!-- 城市导航栏 -->
    </nav>
    <section>
        <!-- 天气预报数据表格 -->
    </section>
</body>
</html>
```

（2）在<nav>标签中定义获取对应城市天气数据的按钮，给按钮绑定 onclick 事件，并将自身的 value 属性作为参数，代码如下：

```html
<nav class="btn">
    <button onclick="load(this.value)" value="北京">北京</button>
    <button onclick="load(this.value)" value="上海">上海</button>
    <button onclick="load(this.value)" value="广州">广州</button>
    <button onclick="load(this.value)" value="深圳">深圳</button>
    <button onclick="load(this.value)" value="武汉">武汉</button>
    <br/><br/>
</nav>
```

（3）编写展示天气数据的表格，代码如下：

```html
<table width="100%">
    <tr>
        <td>城市</td>
        <td>最低气温</td>
        <td>最高气温</td>
        <td>天气</td>
    </tr>
    <tr>
        <td></td>
        <td></td>
        <td></td>
        <td></td>
    </tr>
</table>
```

运行效果如图 15-4 所示。

城市	最低气温	最高气温	天气

图 15-4

步骤 4：制作 CSS 样式

（1）在 index.html 文件中添加<style>标签，在<style>标签中编写页面样式，在头部标签 viewport 中适应移动端页面，设置 content='width=device-width, initial-scale=1.0'。

代码如下：

```html
<!DOCTYPE html>
```

```
<html>
    <head>
        <meta charset="utf-8">
        <meta name='viewport' content='width=device-width, initial-scale=1.0'>
        <title>天气预报</title>
        <style type="text/css">
            /*编写页面样式*/
        </style>
    </head>
    <body>
    </body>
</html>
```

（2）使用 flex 弹性布局使城市导航栏在一行显示。

```
nav {
   display: flex;
   justify-content: space-between; /*各项之间留有等间距的空白*/
   align-items: center;            /*居中对齐弹性盒的各项元素*/
}
```

运行效果如图 15-5 所示。

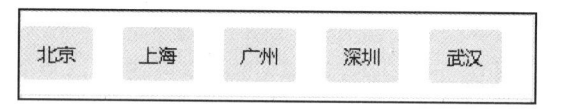

图 15-5

（3）使用 transition 过渡属性给当前被单击的城市添加变宽的效果。

```
.btn button{
    font-size: 0.875rem;
    width: 3.75em;
    height: 2.75em;
    border: 0;
    border-radius: 3px;
    transition: width 100ms;/*在 100ms 内改变 width 属性*/

}
.btn button:active {
    width: 4.5rem;          /*当单击 button 时,宽度变为 4.5rem */
}
```

运行效果如图 15-6 所示。

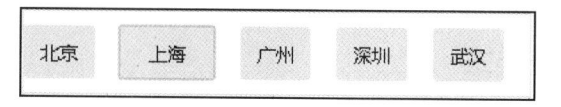

图 15-6

步骤 5：编写 AJAX 请求

1. 在 index.html 中请求 /ListWeather

（1）创建对象：通过判断 window.XMLHttpRequest，创建 XMLHttpRequest 对象。

（2）监听请求状态：为 onreadystatechange 属性设置函数。

（3）判断状态信息和状态码：判断 readyState 和 status 属性，即判断请求是否成功。

（4）创建请求：使用 open( )方法，3 个参数依次为 GET 请求类型、URL 请求路径、true 异步请求。

（5）请求参数：将对应城市的名称通过 URL 传参。

（6）发起请求：send( )方法。

（7）编写 AJAX 请求 XML 数据，代码如下：

```
function load(value){
    var xmlHttp;
    if(window.XMLHttpRequest){
        //使用 IE 7+、Firefox、Chrome、Opera、Safari 浏览器执行代码
        xmlHttp=new XMLHttpRequest();
    }
    xmlHttp.onreadystatechange=function(){
        if (xmlHttp.readyState==4 && xmlHttp.status==200){
            /* 天气信息实时更新 */
        }
    }
    xmlHttp.open("GET", "ListWeather?city="+value, true);
    xmlHttp.send();
}
```

2. 在 index.html 中更新页面

（1）获取 JSON 格式的天气数据：responseText( )方法返回字符串格式，JSON.parse( )方法将字符串解析成 JSON 格式。

（2）获取天气数据并输入<td>中。

（3）获取 JSON 对象中的 name 值并通过 innerHTML 赋值给第 5 个<td>。

（4）获取 JSON 对象中的 min 值并通过 innerHTML 赋值给第 6 个<td>。

（5）获取 JSON 对象中的 max 值并通过 innerHTML 赋值给第 7 个<td>。

（6）获取 JSON 对象中的 weather 值并通过 innerHTML 赋值给第 8 个<td>。

```
var json=JSON.parse(xmlHttp.responseText);
document.getElementsByTagName("td")[4].innerHTML=json.name;
document.getElementsByTagName("td")[5].innerHTML=json.min;
document.getElementsByTagName("td")[6].innerHTML=json.max;
document.getElementsByTagName("td")[7].innerHTML=json.weather;
```

3. 访问页面

访问页面，并单击相应的城市按钮，即可显示天气信息，效果如图 15-1 所示。

参 考 文 献

[1] 北京新奥时代科技有限责任公司．Web前端开发实训案例教程（中级）[M]．北京：电子工业出版社，2019．

[2] 史胜辉，王春明．Java Web框架开发技术（Spring+Spring MVC+MyBatis）[M]．北京：清华大学出版社，2020．